TEACHER'S RESOURCE GUIDE
TO ACCOMPANY

LIVING WITH TECHNOLOGY

Second Edition

 DELMAR PUBLISHERS INC.

NOTICE TO THE READER

Delmar Staff
New Product Acquisitions: Mark W. Huth
Senior Project Editor: Christopher Chien
Production Coordinator: Sandra Woods

Instructional Designer
Margaret Rutherford
Editorial Coordinator
John Hester

For information, address Delmar Publishers Inc.
2 Computer Drive West, Box 15-015
Albany, New York 12212-5015

TABLE OF CONTENTS

PREFACE

This book was developed primarily as a resource book for the textbook *Living With Technology* by Michael Hacker and Robert Barden. However, much of the material can also be used with other Delmar technology texts. It is designed to give the classroom teacher assistance in developing a competency-based technology education curriculum. For each chapter, there is a chapter summary, a chapter outline, a list of major concepts, hints for teaching the chapter, a set of transparency masters, a worksheet (chapter review) with answer key, a chapter test with answer key, a crosstech puzzle, and answers to feedback questions found in the text. Also included are resources where the teacher can obtain more information that might help make the course more productive.

The appendices contain overhead transparency masters, chapter worksheets and answer keys, chapter tests and answer keys, crosstech forms and solution, and a general safety unit.

INTRODUCTION

Over the last fifteen years, we have developed an entirely new vocabulary. Our everyday conversation includes references to pocket calculators, compact discs, genetic engineering, composite materials, lasers, recycling, and space shuttles. What has caused the world to change so radically in so short a time?

Technology (tek nol' e ji) The cause of great societal change. A word that is difficult to define because it surrounds us and takes so many different forms. However we define it, we must agree that it has become a driving force in our society. Pick up any major newspaper; the chances are that one or more of the articles on the front page involve technology. Nuclear power, genetic engineering, acid rain, organ transplants, and "starwars" weapons are just a few newsworthy examples.

Technology has been a part of human life for over a million years, but in recent decades technology has caused more remarkable change to occur than ever before in history. We now are completely surrounded by technological systems. We have become dependent upon the products and services modern technology provides. All industries and all people will continue to be greatly affected by new technological developments. Ours is a highly technological society.

TECHNOLOGY EDUCATION

Although technology can improve the quality of our lives, it can also create undesirable consequences. Many of today's major social issues involve the impacts of technology. Because of the many ways in which technology will continue to affect modern society, the resources, systems, and social impacts related to technological systems must be studied.

Society is developing technological systems so rapidly that it is virtually impossible to predict exactly what kinds of technological skills will be needed in the future. As soon as we have learned about a form of technology, something new takes its place. Therefore, people must develop skills that are useful despite the changes that will no doubt occur.

School is the ideal place to begin understanding technology and the role it plays in our culture. Students should learn that technology is a human endeavor. Whether it is used to benefit or to destroy our society is our decision. Students can learn that people can and must control the development and application of technology.

USING THE TEACHER'S RESOURCE GUIDE

Living With Technology has been written to provide students with the conceptual base necessary to support an activity-based instructional program. The major concepts identified within the chapters can be delivered through the hands-on technology learning activities provided at the end of each chapter. Throughout the text, mathematics, science, and social science concepts are related to the technologies discussed.

A textbook is but one tool in the teacher's tool kit. In systems terms, it is one of the information resources that are combined by the Technology Education process to prepare technologically literate students. As such, the text should not stand on its own without teacher intervention and the use of additional resources. The Teacher's Resource Guide is another tool to be used in conjunction with the second edition of *Living With Technology*. The Instructor's Guide has been incorporated into this edition of the Teacher's Resource Guide.

Living With Technology and the companion Teacher's Resource Guide are particularly valuable assets. The textbook is designed to be read by students to supplement classroom activity. The Teacher's Resource Guide is designed to assist in the preparation for, and delivery of, activity-based instruction. Together they can provide a structure for teachers to deliver content, and can support classroom instruction by providing explanations, examples, related information, activities, and visual illustrations for students.

The textbook can be considered the main source of information, with the other resources used to review and reinforce concepts discussed in the text. Before using the textbook to disseminate information, the teacher may want to refer to the Teacher's Resource Guide for hints on how to teach the unit, and for ideas on preparing the lesson plan.

EXPLORING TECHNOLOGY EDUCATION VIDEOS

The *Exploring Technology Education* videos have been designed to correlate with the material found in *Living With Technology*. The series contains information that covers each of the four main areas of Technology Education; Communication, Construction, Manufacturing, and Energy, Power and Transportation.

The *Exploring Technology Education* videos can be used in conjunction with the textbook and other resources to help give student a look at real-life applications of Technology and how it affects their everyday world. These video programs can be used in addition to the information presented in the *Living With Technology* text and *Living With Technology* Teacher's Resource Guide.

Available with closed captions, the video programs take you to the headquarters of USA Today, where technology has changed how a newspaper is designed and produced, to Australia, the site of the solar-powered car race, to the underwater site of a simulation of space construction, and to construction sites.

VIDEO TAPE PROGRAM SUMMARY:

A. Introduction to Technology

1. Overview of Technology — (15:57)
This video demonstrates the relationship of past to future developments in technology. Students can see the restoration of the Statue of Liberty and Odyssey of the Mind and Rube Goldberg competitions.

2. People, Technology, and the Environment — (13:44)
Technology has its impacts on people and the environment. This video emphasizes that the choices people make have both benefits and risks which involve complex social and environmental problems. An example portrayed in this video is the invention of dynamite by Alfred Nobel, who saw it become an instrument of destruction.

B. Communication

1. Introduction to Communication — (16:58)
This video covers the history of communication technology. People are shown using simple to complex communication technologies. The systems model (input, process, output, feedback) is shown in a communication system. Students can glimpse into the future via a visit by video to MIT's Media Lab.

2. Designing Messages — (13:18)

Students are taken to USA Today by video and shown how a newspaper is designed using three considerations for designing a message — audience, content, and form. The video shows the elements of good design. The impact of computerization of the newspaper is explored.

3. Producing and Transmitting Messages — (16:29)

This video further develops the video Designing Messages by again returning students to USA Today to explore the use of computers in preparing and transmitting information. Students view modernized typesetting, laser scanning for color pictures, and transmissions by facsimile scanners in the simultaneous printing of the newspaper across the country. The worldwide broadcast of the Live Aid concert by satellite is also shown.

4. Evaluating Messages — (12:23)

This video shows students the importance of feedback. Students view footage of a plane colliding with the Empire State Building, the Apollo 13 flight, and use of modems. They see how overnight express service and TV and radio ratings provide a rapid form of feedback.

C. Construction

1. Introduction to Construction — (15:32)

Students are shown types of construction, development of construction, and its relationship to society and the environment. Residential, civil, and commercial construction projects are shown. The video shows how the systems approach is applied when it takes students to a middle school and airport. Construction in space is covered when students view other students performing construction tasks underwater. The use of robots in construction on the moon and Mars is explored.

2. Designing and Planning a Structure — (12:42)

This video emphasizes that design and planning processes in construction are very important. It shows how these processes affect the outcome of construction of bridges such as the Golden Gate Bridge and the Tacoma Narrows Bridge. The video then shows an application of the bridge building concept depicting students using a systems approach to building model bridges. Construction underwater is again approached when students attempt to construct a geodesic dome underwater in a gravity-free environment.

3. Building Safely — (13:47)

Students view the steps in building a house: surveying and preparing the site, building the footing, laying down joists, subflooring, and flooring, and framing walls. Materials and methods for construction in the future are also covered.

4. Finishing a Structure — (13:33)

This program is a continuation of video #3 (Building Safely). Students are shown installation of the subsystems of a structure (heating, cooling, plumbing, and electrical). The students are introduced to the processes, materials, and techniques required to finish the structure. They are also shown a new concept in construction, the "Smart House."

D. Manufacturing

1. Introduction to Manufacturing — (15:11)

The evolution, development, and impact of manufacturing technology is illustrated in this video. Students are shown Henry Ford's assembly line. The universal systems model is related to the production of the Model T Ford. Students are shown the use of robotics in manufacturing and the making of compact discs.

2. Manufacturing Systems — (13:03)

Intermittent, continuous, and custom manufacturing systems and their essential elements are introduced in this video program. Examples of each type of the manufacturing systems are shown. The four elements of manufacturing — finances, material, people, and facilities — are explored. Using the example of the Apple Computer Company, students are shown how the system model is used in manufacturing.

3. **Using and Evaluating Materials** — (12:50)

The right choice of materials for each project is discussed in this video program. The properties of materials such as wood, metals, brick, ceramics, plastics, and composites are explained. Testing of materials is shown in a visit to Batalle Laboratories. Students are then asked to compare properties of aluminum and wooden baseball bats.

4. **Manufacturing Process** — (12:15)

The six manufacturing processes—separating, cutting, forming, casting, conditioning, assembling, and finishing—are introduced. Students are taken by video to a steel foundry to view these processes and the upgrading and refinement of these processes. The use of CAD and laser sculpting show students updated applications of manufacturing processes.

5. **Manufacturing Process Planning** — (14:11)

Quality control, productivity, and new products involve constant planning. Students view these planning processes in a visit to a B.F. Goodrich plant. Students also visit by video a Bobcat vehicle plant and a greeting card publisher who must make use of planning processes in the manufacture of their products.

E. *Energy, Power, and Transportation*

1. **Overview of Energy** — (15:27)

Energy is defined in this video program. Energy sources, forms, technical advances, uses, and impacts on society and the environment are covered. Students are told about exhaustible, renewable, and inexhaustible sources of energy. Students learn about environmental problems created by use of fossil fuels and of the attempts to harness inexhaustible forms of energy.

2. **Conversion of Energy into Power** — (14:32)

Conversion of one form of energy into another is covered in this video program. Thermal, chemical, mechanical, radiant, electrical, and nuclear energy are explained. Students are able to view a coal-fired electrical plant, a solar-powered car, and a human-powered aircraft. The need for efficient use of energy, including the use of nuclear fission from seawater, is discussed.

3. **Transmission, Control, and Storage of Power** — (12:17)

The control, transmission, and storage of power are discussed in this video. Students can see how other students construct and attempt to control a Hovercraft. A model blimp and fighter planes on an aircraft carrier illustrate the four storage and transmission systems—mechanical, fluid, electrical and thermal. The storage of power by these systems is discussed.

4. **Transportation Systems** — (12:49)

The history and growth and kinds of transportation systems are portrayed in this video program. Students can view transportation systems used on the earth, in water, in the air, and in space. An application of intermodal transportation, the Federal Express, is shown. The possibility of solar-powered cars, magnetic-levitation trains, and orbiting space stations is explored.

LESSON PLANNING

Lesson planning is a critical and integral part of delivering effective instruction. After reviewing each chapter you might choose to develop a plan to facilitate the instructional process. The Teacher's Resource Guide contains a lesson planning guide and sample lesson plan. The sample lesson plan is provided as an example of the process of preparing and planning for an instructional unit. The lesson planning guide is provided to assist in collecting and organizing ideas, thoughts, and resources for each instructional unit. Each chapter is condensed into a chapter summary, with an outline and the major concepts listed to facilitate the lesson planning process.

Since there is no one perfect method for preparing an instructional plan, the Teacher's Resource Guide was designed to supply a great deal of information in an easy-to-use format.

The first part of the Teacher's Resource Guide provides material which can be used in lesson planning. The second part of the Teacher's Resource Guide, the appendices, provides support material for the delivery and evaluation of each lesson.

The lesson planning features of the Teacher's Resource Guide include:

Chapter Summary

A brief overview of each chapter is provided in the Teacher's Resource Guide to quickly help the instructor become familiar with the content of each chapter.

Chapter Outline

Chapter outlines, with textbook pages identified, are provided to enable the instructor to locate and review, at a glance, most of the major concepts in the chapter. Chapter outlines are provided to assist the instructor in correlating the textbook content to desired instruction. The outlines have been developed according to the sequence of content within the chapter.

Major Concepts

The major concepts for each chapter, found in the Teacher's Resource Guide, are the same ones found in the beginning of each chapter in the textbook. You will find the major concepts on an overhead transparency master. The major concepts can be used as an anticipatory set or focus for the chapter. In addition, major concepts can be reworded into behavioral objectives.

Hints

The Teacher's Resource Guide also contains specific "hints" to help the teacher relay to the students the material covered in the chapter. Hints to the teacher include additional activity ideas, as well as pertinent information related to safety. It also contains "hints" for helping the teacher complete the activities that are given at the end of each chapter.

The lesson support material provided in the Teacher's Resource Guide includes:

Overhead Transparency Masters

To assist in the delivery of each lesson, overhead transparency masters are provided for each chapter. Overhead transparencies can play a vital role in the student's ability to retain information. The students not only hear, but also see the information being presented. The first overhead transparency master covers the MAJOR CONCEPTS of each chapter. The remaining masters contain information that reinforces the material in each chapter. As you make the transparencies, they can be put into a binder for future use. If the master is used on an opaque projector and enlarged on poster board or similar material, the students can have a daily visual reinforcement of the material.

Use the transparencies to introduce the chapter, discuss the major ideas in the chapter, summarize materials covered, and/or to reteach information. As you use the overhead projector, you might try stepping to the side as you speak; let the transparency remain on the overhead projector so that the students can see, as well as hear, what you are saying. When you are finished, put the transparency into a binder for safekeeping.

Chapter Review Questions and Answers

Chapter review questions and answers are provided as an aid to the teacher. Some questions require different levels of thinking skill. The questions range from knowledge, comprehension, and application questions (lower levels of Bloom's taxonomy) to analysis, synthesis, and evaluation question (requiring higher level skills). You may want to choose Feedback questions that are appropriate to the level of the students.

Crosstech Puzzles

The Crosstech Puzzles cover the vocabulary words for each chapter. These puzzles were designed to complement the chapter reviews in the Teacher's Resource Guide. Solutions are provided as an aid to the teacher.

Chapter Worksheets and Answer Keys

The chapter worksheets (reviews) are written with the major concepts in mind. Questions are in the form of Multiple Choice, True/False, Matching, Identification, and Fill in the Blanks. The worksheets are designed to comprehensively question the student's knowledge on all major ideas of each chapter. Answer sheets are included for the convenience of the instructor. The value of each question depends on the number of answers on each worksheet. Corrected worksheets can be used as a test review.

Chapter Tests and Answer Keys

The tests are designed to evaluate student comprehension of the major ideas or information of each chapter. Answers to test questions are to be placed in blanks on the left side of the question. Answer sheets are included for the convenience of the instructor.

General Safety

Safety is an important aspect of teaching about technology; therefore it should be covered in each lesson as part of the content. In addition, safety should be covered as a unit by itself to help stress its significance. Appendix F has been developed as a resource to help prepare for and provide safety instruction. It includes general safety rules, which should be tailored for each classroom setting, a list of safety resources, and a general safety test.

USING THE SPECIAL FEATURES IN THE TEXT

Living With Technology includes some unique features to enhance interest and to assist students in becoming technologically literate.

Major Concepts. These are provided at the beginning of each chapter. These major concepts prepare the student to be aware of key ideas while reading. Each major concept is repeated in the margin adjacent to its related discussion in the text. It should be noted by teachers that some of the major concepts may be a bit complex for students at first reading. Although the text will eventually explain each clearly, some students may become intimidated by them. It is therefore suggested, that the teacher preview the major concepts, identify those which will relate to the particular reading assignment, and explain them briefly in order to raise the students' comfort level.

Key Terms. These are highlighted the first time they appear in the text. Students will find many of these terms to be new to them. They should be reminded that definitions are found in the text glossary. Again, teachers may wish to intervene by providing some discussion relative to certain terminology.

Connections. These are sections within each chapter which describe the relationships between Technology and Math/Science/Social Studies and the information that is being presented in the chapter. These sections also allow students to become aware of the many ways technology encompasses their everyday routines, and it makes the students aware of real-life applications of technology.

Crosstech Puzzles. These provide a motivating review of key terms and reinforce major concepts. The crosstech puzzle masters can be duplicated for student use. The crosstech puzzles can be used as homework assignments, quizzes, or substitute teacher lessons.

Chapter Summaries. These have been carefully written so that major concepts are revisited and reviewed. Students should be encouraged to read the portion of the summary that pertains to the specific reading assignment, and then asked to reread the entire summary after completing the reading of the chapter.

Activities. These are at the end of each chapter for the student to work on, and relate to the material in the chapter. These activities are divided into six sections. The first section is *Problem Situation*, which describes a problem or need that exists. The next section, *Design Brief*, offers a short description of an activity or problem that the student is to solve. The third section, *Procedure*, lists the method by which to complete the Design Brief. The fourth and fifth sections are related to technology and its application in the classroom, as well as real-life applications. The sixth section, *Summing Up*, gives the students the opportunity to discuss the material they have just covered.

A *Technology Time Line* is included for historical reference. Frequently, technology teachers request that students model technological devices and place them on an historical time line. Students will find the Time Line to be a useful resource which illustrates how technology has grown exponentially since prehistoric times.

OPTIMIZING THE BENEFITS

In using *Living With Technology*, it is suggested that teachers first present a short lesson which sensitizes the students to the particular topic of study. The major concepts that the students are expected to learn should be highlighted in the lesson. The teacher may wish to draw examples from both the technology learning activity and the material in the reading to be assigned to help the students grasp the key points. New terminology should be introduced and defined so that difficult new words will not deter the students from continuing to read. Only after students are introduced to the new material should the reading be assigned.

Once the students have completed the reading assignment, the teacher should take a few minutes of class time to review the textual material, again highlighting the major concepts. To reinforce the reading, the teacher might wish to provide some additional examples or draw them from the students' own experiences.

The text has been written to provide maximum reinforcement of those major concepts identified at the beginning of the chapters. These concepts are reprinted in the margin adjacent to the discussion in the text. The chapter summaries also refer to the major concepts. Therefore, the teacher might wish to assign the students to read that portion of the summary that deals with the specific 5–8 page reading. At the conclusion of the chapter, the entire summary could be reread.

Each chapter concludes with a series of questions (Feedback) which directly relate to the major concepts. These questions are of various types (recall, analysis, synthesis) and should be selectively assigned by the teacher based upon the academic ability of the class. The Crosstech puzzle in each chapter serves as an excellent tool to teach the key terms. Finally, at the end of each chapter are activities that would be appropriate for conveying the major concepts identified in the chapter.

The suggested method for using the text can be summarized as follows:

1. Sensitize class to topic by presenting related lesson.
2. Emphasize major concepts that students will encounter in the reading. Link these concepts to the technology learning activity.
3. Define new terminology.
4. Assign 5–8 pages of reading and proper part of summary as homework.
5. Review the reading assignment during class.
6. Continue to assign 5–8-page reading assignments and review in class. After students have concluded the chapter, have them reread the summary.
7. Assign selected Feedback questions.
8. Assign Crosstech puzzle.
9. Complete technology learning activity (this should be done throughout the reading of the chapter).
10. Review all assignments (Feedback questions, Crosstech, and technology learning activity).

Length of Reading Assignments

Reading should be assigned on a regular basis. Students will learn to expect that the study of technology includes work outside of class. It is suggested that the reading assignments be short and frequent. For middle and junior high students, 5–8 pages per sitting is recommended, with a maximum of 10 pages being assigned at any one time. If we assume that a typical Introduction to Technology course is comprised of 36 weeks, then a minimum of 13 pages per week would permit students to complete the entire text. Generally this would involve two readings per week. Chapters are approximately 25–35 pages in length; therefore, two to three weeks would be devoted to the content in one chapter.

Integrating Readings With Hands-On Activity

If two or three weeks are spent on each chapter, technology learning activities can be designed to fit that time frame; or they may be longer and span the content and time of several chapters. Technology Education is an activity-based program. Ideally the reading and the technology learning activities will act in harmony to solidify the major concepts in the minds of students. In order to arouse student interest and keep motivation high, activities should be commenced shortly after the readings are assigned.

The reading can serve as an introduction to the activity. For example, assume that students are learning about the evolution of technology, the topic covered by Chapter 1 in the text. In an introductory lesson, the Technology teacher would present major concepts identified within Chapter 1. The teacher might address a major concept such as "people create technological devices and systems to satisfy basic needs and wants" by eliciting examples from students of devices and systems created to satisfy basic human needs and wants. The major concept could be easily linked to the development of tools or time-keeping devices. A typical activity, therefore, might involve the design and construction of a primitive tool or an early time-keeping device. Continued reading would provide other examples.

The activity is a vehicle through which concepts are taught. As additional readings are assigned, teachers may wish to stress each major concept at the time it best relates to the chosen activity. In Technology Education programs, there will be continual interplay between text, teacher, and activity.

INTEGRATED INSTRUCTION

In *Living With Technology*, processes central to industrial technology (communication, manufacturing, construction, and transportation) are fully discussed. Chapters devoted to the evolution of technology, problem solving in technology, electronics and computers, biotechnical systems, and the impact and future of technology round out this comprehensive study.

The text uses a systems approach in describing technologies. Through the use of the basic system model, students are provided with a conceptual tool with which technologies can be analyzed. Students will be able to break down and understand technologies not previously encountered, as well as emerging technologies that have not been described in school. The system analogy can also be used to describe and understand other complexities of modern life, including economic, political, and education systems.

Through the systems approach the text conceives all technologies as systems with reoccurring elements. These elements (inputs, resources, processes, outputs, feedback, and control) are common to all technological systems, and therefore provide a basis for an organized study of technology as a discipline.

To be understood and used to advantage, technology must be viewed within its societal context. It is no longer enough to simply be familiar with the technical process itself. The careful choice of resources to be combined by the technological process and the assessment of possible outputs are keys to productivity. Citizens who are able to make constructive contributions to society will be those who can make decisions based upon an understanding of the interaction of technological, economic, cultural, and environmental systems.

EVALUATION

When the time comes to evaluate the students' performance, the text and supplementary materials enable teachers to have several options available to them for evaluating the students' performance. First the teacher should have a guideline of intended outcomes for the course.

Once teachers have the guidelines or intended outcomes for the course, they may then begin to look at their options for evaluation, and choose the ones that best fit their needs.

Short answer and essay questions can be a valuable tool in evaluating student performance. The Feedback questions at the end of each chapter, along with the Crosstech puzzle, can serve this purpose. This enables the teacher to check for reading comprehension and how much students understand the material they read. Essay questions should be phrased so that the teacher can assess what the student understands about the topic.

The student projects and activities listed in the book and explained in the Teacher's Resource Guide are another good way in which evaluation of the students' performance can be made. Evaluation of the activity or project could be broken down into parts where the student is perhaps evaluated on the following things; following directions, completion of project or activity, effort made, neatness, accuracy, etc. Teachers could develop their own criteria to meet their needs.

The students themselves could be asked or required to do their own evaluation of how well they are performing in class. This could be done by either self-evaluation, or evaluation by ones' peers. If the class is doing the evaluation, they could be set up to resemble a Board of Directors, or perhaps even Patent Office officials, who could develop their own criteria for evaluation of their classmates' performance. It is important for the teacher to stress non-biased judgement, or evaluation. This is also a good time for the teacher to explain to the class what a value judgement is and how the students' feelings will in fact alter their evaluation.

Design Folders or portfolios could also be used as an evaluation tool. As explained in Chapter 2, problem-solving requires the student to keep a record of all data that was gathered. A design folder enables the student to go about his or her problem-solving in a uniform fashion, and provides a place where all materials—drawings, sketches, brainstorming activities, etc.—may be kept. When evaluating the students' performance, the teacher may use the design folder as a tool to see what the students have done in order to accomplish the goals that have been set out for them.

SAMPLE MONTHLY PLAN

This plan exemplifies the integration of lecture, demonstration, student activity, and reading assignments. The plan is correlated with Chapter 13, "Controlling the System".

Day 1: Sensitize class to topic by presenting related lesson entitled "How We Control Technology". Discuss familiar examples of controlled systems. There are many controlled systems in the human body that can be used as examples, such as temperature control (shivering and perspiring), and control of adrenalin (during reaction to an emergency, adrenalin carried in the blood makes your organs, muscles, and nervous system work harder.) Discuss open and closed loop systems. Emphasize major concepts 1-2. Assign reading, Chapter 13, pages 362-365.

Day 2: Review reading, highlighting major concepts 1-2, and key terms. Elicit additional examples of open- and closed-loop systems from students. Introduce the components in the feedback loop (sensors, decision makers, and controllers). Introduce hands-on activity. Activity should reinforce system control concepts. One idea is to build a tantalizer by simulation of a production line. The tantalizer is a toy which uses a mirror to teach the concept of feedback control.

Day 3: Contrast control of mass production systems with craft production systems. Indicate differences between human control and machine control. Discuss instances of open-loop (i.e., timers and computer programs), and feedback control of automated systems.

Day 4: Discuss flow chart, work stations, and organization of the production line. Discuss use of jigs and fixtures. Emphasize major concepts 3-5. Assign reading, Chapter 13, pages 365-369.

Day 5: Review reading, highlighting major concepts 3-5, and key terms. Have students precut, prepare and store necessary materials of tantalizer.

Days 6 and 7: Train students to operate production line. Assign individual tasks. Emphasize major concept 6 by indicating how the production system is designed to produce a particular set of outputs (a given number of games of a given quality, in a given period of time). Discuss sensors and point out sensing devices necessary to the production process (both human and technological sensors). Assign reading, Chapter 13, pages 369-373.

Days 8 and 9: Review procedures for production. Review major concept 6, eliciting from class the desired outputs and how the system is designed to produce them. Run the actual production. During the production run, collect data about the time it takes to carry out each task. Later, this data will be analyzed to determine how many workers should be assigned to each station, and how bottlenecks can be eliminated. Implement quality control procedures for each operation and for the final product. Observe and record data about quality on a check-off sheet.

Day 10: Provide instructions for use of tantalizer. Instructions can be written and printed by students. Try the tantalizer game in class, focusing on the need for monitoring, decision making, and control in accurately tracing the object. In the human feedback loop, the monitoring is done by the eye, the decision making by the brain, and controlling by the brain and the hand. Explain how decision makers are monitors. They compare the actual results with the desired results. Explain how controllers are used to turn the process on or off, or to adjust it in some way. Assign reading, Chapter 13, pages 374-380.

Day 11: Debrief with students on the production run. Analyze time and quality control data. Reinforce mathematical concepts by graphing time vs. product quantity by drawing a histogram (frequency distribution). Draw the time on the horizontal axis in five-minute intervals. On the vertical axis, represent the number of items produced during each five-minute time period. `The histogram is used to show production levels over a given period of time. Determine which tasks needed a greater or lesser number of people. Reinforce science concepts by pointing out how images are reflected, and how the angle of incidence equals the angle of reflection; introduce the concept of vectors. Discuss problems with feedback. The feedback students received was conflicting feedback. Relate the mirror activity to human-to-human feedback, and the signals we get from people (teacher to student, student to student). Discuss delayed feedback. Draw a wavy line on the blackboard. Have a student in front of it on a dolly that is to be pulled along the floor. Blindfold the student. Ask the class, or a class member, to call out instructions to the student as he or she tries to retrace the wavy line without seeing it. The feedback for the class is delayed, and the student will be unable to be perfectly accurate.

Day 12: Lesson on sensors. How could sensors have been used to improve the control of any of the manufacturing processes? Relate common instances of sensing such as optical sensing (automatic supermarket doors); infrared sensing (autofocus cameras); magnetic sensing (burglar alarms on windows); and thermal sensing (thermocouple on kilns and ovens). Assign reading, Chapter 13, pages 380–384.

Day 13: Students experiment with a variety of systems using electrical, electronic, mechanical, optical, thermal, and magnetic sensors. Assign Feedback questions 1–4.

Day 14: Introduce hands-on activity, construction of a device to control house lights at dusk. Compare feedback control from sensors to control via a program (exemplified by a timer). How could a light control device function with open-loop (program) control? What are the disadvantages of using a time to control lights in the home? Demonstrate proper and safe construction techniques, and use of tools.

Day 15: Discuss answers to Feedback questions 1–4. Students begin construction of light-control device. Obtain schematic diagram and parts list from 200 in 1 Electronics Kit, Tandy Corporation, Fort Worth, TX 76102, or similar electronic project source.

Days 16–19: Continue construction. Assign feedback questions 5–7. Assign crosstech as homework.

Day 20: Complete construction. Discuss answers to Feedback questions 5–7. Hand out answers to Crosstech and discuss.

LESSON PLAN WORKSHEET

NAME _____ DATE _____

PERIODS _____ GRADE LEVEL _____ CHAPTER _____

OBJECTIVES:

_____

_____

_____

_____

RESOURCES: MATERIALS REQUIRED:

_____   _____

_____   _____

_____   _____

_____   _____

METHODS OF EVALUATION: MODIFICATIONS: SPECIAL NEEDS

_____   _____

_____   _____

_____   _____

_____   _____

MULTI-DISCIPLINARY STRATEGIES:

_____

_____

_____

_____

PROCEDURES TO BE FOLLOWED:

_____

_____

_____

_____

MODIFICATIONS TO BE MADE AFTER ASSESSMENT OF STUDENT RESPONSE:

_____

_____

_____

MODIFICATION WORKSHEET

Modified Activity: _____

OBJECTIVES:

1. _____

2. _____

3. _____

4. _____

RESOURCES

People: Who will be involved in the activity?

_____

_____

_____

_____

Information: Learning material

_____

_____

_____

_____

_____

Materials: Supplies, etc.

_____

_____

_____

_____

_____

Tools and Machines:

_____

_____

_____

_____

_____

Energy Sources: Kinds needed

_____

_____

_____

_____

_____

Time: Length of activity

_____

_____

_____

Capital: Amount to be budgeted

_____

_____

_____

TECHNOLOGY IN A CHANGING WORLD

Chapter 1, Text pages 1 to 25

Chapter Summary

Chapter 1, "Technology In A Changing World", defines technology and its relationship to science. The chapter addresses the reasons, importance, and benefits of technology.

Chapter Outline

A. What is the Difference between Science and Technology? (p. 4)
B. Why Study Technology? (p. 6)
C. How Does Technology Affect our Lives? (pp. 7–10)
   1. How Technology Influences our Routines
   2. How Technology Satisfies our Needs
      a. Agriculture
      b. Medicine
      c. Production
      d. Energy
      e. Communication
      f. Transportation
D. When Did Technology Begin? (p. 6)
   1. The Stone Age
   2. The Bronze Age
   3. The Iron Age
E. The Industrial Revolution (p. 12)
F. Exponential Change (p. 14)
   1. Technological Time Lines
G. The Three Technological Eras (p. 18)
   1. Agricultural Era
   2. Industrial Era
   3. Information Age
H. Technological Literacy (p. 19)
I. Summary (p. 20)
J. Review Questions (p. 21)
K. Crosstech
L. Activities (pp. 22–25)

Major Concepts:

- Technology affects our routines.
- Science is the study of why natural things happen the way they do.
- Technology is the use of knowledge to turn resources into goods and services that society needs.
- Science and technology affect all people.
- People create technological devices and systems to satisfy basic needs and wants.
- Technology is responsible for a great deal of the progress of the human race.
- Technology can create both positive and negative social outcomes.
- Combining simple technologies can create newer and more powerful technologies.
- Technology has existed since the beginning of the human race, but it is growing at a faster rate today than ever before.

Student Behavioral Objectives:

Upon completion of the "Now and Then" activity, the student should be able to:

1. Explain the historical progression of technological developments.
2. Give examples of technological devices used in the various periods of history.
3. Distinguish between two historical time periods of technological development by sketching examples of the technologies of the two times.

Hints for Chapter 1

Instructional Suggestions:

a. Encourage students to bring magazines related to technology
b. Schedule trips to the library for research
c. Schedule time for viewing of related videotapes
d. Collect materials ahead of time for development of timeline.

On page 4, the words "scientists" and "technologists" are differentiated. A teacher-directed discussion could be held to reinforce the differences. Students might be asked whether they would like to study scientific relationships (as the scientist), or use scientific knowledge to build things (as the technologist), and why. For further reinforcement, there is a chart on page 5 that compares the two words.

A mathematical concept central to technological development is that of exponential growth. To explain the concept, the following graphs can be related to the cassette tape example presented on pages 14–16. These may prove useful to describe the generation of an exponential curve.

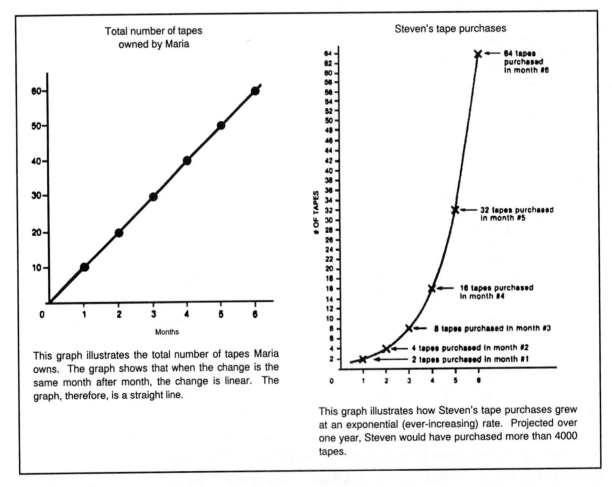

Total number of tapes owned by Maria

This graph illustrates the total number of tapes Maria owns. The graph shows that when the change is the same month after month, the change is linear. The graph, therefore, is a straight line.

Steven's tape purchases

64 tapes purchased in month #6

32 tapes purchased in month #5

16 tapes purchased in month #4

8 tapes purchased in month #3

4 tapes purchased in month #2

2 tapes purchased in month #1

This graph illustrates how Steven's tape purchases grew at an exponential (ever-increasing) rate. Projected over one year, Steven would have purchased more than 4000 tapes.

A time line, an important graphic aid, is given on page 17. The time line is a type of graph. It shows how much change has occurred over a given period of time. A point to stress is that certain events led to a rapid increase in technological development. For example, it should be noted that the development of writing (see page 17) greatly accelerated technological change.

In discussing technological literacy, several dimensions can be pointed out, including the ability to create, use, control, and assess the impacts of technology.

An activity that might cement the idea of exponential growth is the generation of a time line of technological developments. Students might be tasked with drawing a picture or constructing a 3-D model of a device, and placing it on a time line. Toward the end of the time line (representing modern times), the devices will be crowded, indicating the increasingly rapid rate of technological change.

Another interesting activity might involve an international exchange. Through the United Nations or a foreign embassy, the teacher may arrange to pair his or her school with one in a foreign country. Through fundraising or contributions, the United States students might provide a camera and film to the foreign students. Each class would be asked to take photographs of the kinds of technology they encounter over a period of one week. The Technology students would then develop and print the film. Comparing the photographs can help clarify how technology influences people's routines, and how technology is responsible for a great deal of the progress of the human race.

Hints for Activities

1. Interview individuals over the age of 65.
2. Divide the class into teams and have each team come up with ten predictions for the future. A spokesperson of each group reports the predictions. One person should keep a tabulation of those predictions which were most frequent.
3. Assign each student a period of time. Given parameters, have each student complete a poster of his or her period. Mount these posters down a hall or around the room.

Individual Activity:

Assign students an outside-of-class activity:
   a. Interview individuals over the age of 60
      ■ Use a standardized interview sheet developed by the class
      ■ Interview individuals and complete interview sheets
   b. Record—by sketching or photographing—tools of the past
   c. Submit the findings and report them to the class

Cooperative Activity:

Divide students into teams
   a. Assign teams the task of predicting the future in technology
   b. Teams brainstorm, sketch, and report findings to class
   c. Teams meet again for possible extensions and solutions to one of the other team's predictions
   d. Meet as a class to discuss and justify solutions

Cooperative Activity:

Divide students into teams:
   a. Assign each group a period of time
   b. Given specifications for size of drawing, have each team sketch and paint technological developments of that time
   c. Assemble the timeline and mount on walls
   d. Invite math and science teachers to judge panels
   e. Publicize and invite all students to view the timeline development

Overhead Transparency Masters can be found in Appendix A.

Chapter Review Questions and Answers can be found in Appendix B.

Crosstech Puzzles and Solutions can be found in Appendix C.

Chapter Evaluation Material can be found in Appendix D.

Technology Activity Guidelines can be found in Appendix E.

A General Safety Unit can be found in Appendix F.

RESOURCES FOR TECHNOLOGY

Chapter 2, Text pages 26 to 43

Chapter Summary

Chapter 2, "Resources For Technology", covers the seven types of resources used in technological systems: people, information, materials, tools and machines, energy, capital, and time. Renewable and nonrenewable sources of energy are discussed.

Chapter Outline

A. Technological Resources (p. 27)
B. People (p. 28)
C. Information (p. 28)
D. Materials (p. 29)
    1. Raw Materials
        a. Renewable and Nonrenewable Raw Materials
    2. Limited and Unlimited Resources
    3. Synthetic Materials
E. Tools and Machines (p. 31)
    1. Hand Tools
    2. Machines
    3. Electronic Tools and Machines
    4. Six Simple Machines
    5. Optical Tools
F. Energy (p. 34)
    1. Limited, unlimited, and renewable energy sources
G. Capital (p. 35)
H. Time (p. 36)
    1. Time Measurement Throughout the Ages
I. Summary (p. 36)
J. Review Questions (p. 39)
K. Crosstech
L. Activities (pp. 40–43)

Major Concepts:

■ Every technological system makes use of seven types of resources.
■ The seven resources used in technological systems are people, information, materials, tools and machines, energy, capital, and time.
■ Since there is a limited amount of certain resources on the earth, we must use resources wisely.
■ Solving technological problems requires skill in using all seven resources.

Student Behavioral Objectives:

Upon completion of the "It's About Time" activity, the student should be able to:

1. Explain what is happening to time as a resource.
2. Give examples of time devices used through history.
3. Design and construct a battery-operated quartz movement.
4. Demonstrate the safe use of tools and machines.
5. Develop a CAD drawing of a battery-operated quartz movement.

Hints for Chapter 2

Instructional Suggestions:

    a. Announce in advance the activity, materials needed, and outside-of-class assignments.

b. Encourage students to bring pictures and/or articles about time devices to be displayed on a bulletin board.

c. Schedule a guest speaker to come and talk to the class about time studies done to improve productivity.

d. Have different classes construct the battery-operated quartz movement using both individual and mass production techniques. Assign a time to complete time studies on each phase of construction. Have the students report the results of the time study to the class. Ask how this impacts production in industry. The focus of instruction related to this chapter would be to impress the students with the idea that every technological system makes use of seven types of resources. A trap to avoid would be to overstress materials and tools and machines as resources and neglect others (people, information, energy, capital, and time).

Depending upon the academic level of the class, teachers may wish to expand upon terms which are presented in the text but not detailed. Activities could include the creation of a product with specified limits on the use of certain resources. For example, students might be restricted to the use of only certain information resources. The teacher might not be an available source of information during this experience. Similarly, an energy resource, like electricity, may be restricted. Students could be asked to explain how better access to these resources might have improved results.

A Technology T-shirt could be produced with a logo symbolizing the seven resources. The logo could be computer generated and photographically silk-screened.

Have students use different colored pins for various kinds of resources and indicate on a map of the world where sources of these resources are. Point out how the world is interdependent, and that the United States relies on other countries for some of its strategic resources.

Choose a resource like energy or forest resources. Have students research the increase or decrease in known reserves over the past 10 years in a particular geographic area and plot the supply versus time on a graph. A good reference for this information is *The Resourceful Earth* by Julian L. Simon and Herman Khan, Basil Blackwell Inc., Publishers, 432 Park Avenue South, Suite 1505, New York City, NY 10016.

Hints for Activities

1. Have teams brainstorm and sketch their ideas of innovative time devices.
2. Design and construct models of ancient time devices.
3. Have students collect broken clocks and watches and dissect or tear down the pieces to see what is inside and how it works. Also have them try fixing the time pieces.

Individual Activity:

Assign an out-of-class activity.

a. Students look in magazines, newspapers, and brochures for pictures of traditional, conventional, and space-age time devices.

b. Have students assemble a notebook or use these for a bulletin board display.

c. Select students to report on their findings about time pieces throughout history.

Cooperative Activity:

Assign teams to research specific types of time devices such as water, mechanical, quartz movement, and atomic clocks:

a. Teams go to libraries, view videos, and visit stores to get literature and take pictures of time devices.

b. Each team prepares a presentation of the findings of their research.

c. Each team presents to the class a videotape of their presentation.

d. Have the class as a group share other information learned about time during their research.

Overhead Transparency Masters can be found in Appendix A.

Chapter Review Questions and Answers can be found in Appendix B.

Crosstech Puzzles and Solutions can be found in Appendix C.

Chapter Evaluation Material can be found in Appendix D.

Technology Activity Guidelines can be found in Appendix E.

A General Safety Unit can be found in Appendix F.

PROBLEM SOLVING AND SYSTEMS

Chapter 3, Text pages 44 to 86

Chapter Summary

Chapter 3, "Problem Solving and Systems", deals with solving problems, systems, and issues evolving from problem solving. The parts of a system such as input, process, output, and feedback are explained.

Chapter Outline

A. Problem Solving (p. 45)
B. Good Design in Problem Solving (p. 46)
C. Examples of Well-Designed Technological Solutions (p. 46)
D. The Technological Method of Problem Solving (pp. 49–60)
E. Solving Real-World Problems (p. 60)
F. Systems (p. 62)
G. The Basic System Model (p. 62)
   1. Inputs
   2. The Process
   3. Outputs
   4. Feedback
G. Multiple Outputs (p. 68)
H. Subsystems (p. 68)
I. Summary (p. 70)
J. Review Questions (p. 72)
K. Crosstech
L. Activities (pp. 74–85)

Major Concepts:

■ People must solve problems that involve the environment, society, and the individual.

■ A carefully thought out multi-step procedure is the best way to solve problems.

■ A good solution often requires making trade-offs.

■ Technological decisions must take both human needs and the protection of the environment into consideration.

■ People design technological systems to satisfy human needs and wants.

■ All systems have inputs, a process, and outputs.

■ The basic system model can be used to analyze all kinds of systems.

■ Feedback is used to make the actual result of a system come as close as possible to the desired result.

■ Systems often have several outputs, some of which may be undesirable.

■ Subsystems can be combined to produce more powerful systems.

Student Behavioral Objectives:

Upon completion of the "Model Rocket-Powered Spacecraft" activity, the student should be able to:

1. Explain the systems and subsystems of a rocket-powered spacecraft launch.

2. Follow a procedure developed by the student in the design and construction of a rocket.

3. Demonstrate the safe use of tools and machines in the construction of a rocket.

Hints for Chapter 3

Instructional Suggestions:

a. Announce in advance the activity, materials needed, and outside-of-class assignments.
b. Encourage students to bring pictures and/or articles about spacecraft launches. Have them choose one launch and diagram a systems approach of the launch.
c. Order videotapes related to spacecraft launches and show these.
d. Set up a date for a launch time. Invite other classes, parents, and the news media to view the launches.

Before students can truly understand a system and a system model and how it can be applied to Technology, they must first understand the need and procedure of problem solving. The steps of a typical problem-solving process are shown on pages 49–60; each step is related to a portion of the system model. Using the same model that is used to conceptualize various technologies, the students thus have a tool to help in solving new problems. The concept that feedback is used in problem solving is important because it suggests that the first solution tried is not necessarily the best, and that people can learn from failures. A very effective way to teach these concepts is through a design problem, such as the one suggested in this section, that gives the students freedom to choose and implement solutions to a problem. Problems such as this also show that a single problem can have many correct solutions, although there is often one best solution.

In order to promote the skills of investigation, the teacher might wish to ascertain that the student can:

a. recognize the existence of a problem that could be solved through technical means
b. follow the seven steps in the problem-solving model shown on page 49
c. use information resources and judge how relevant they are
d. synthesize knowledge in order to reach conclusions

The problem should therefore, lend itself to the development of investigatory skills. For example, assume the problem is make an original map of the area immediately surrounding the school. A possible approach for the teacher to use is:

a. Help the student recognize that the problem can be solved through technical means. Elicit from the students the desirability of viewing the school and surrounding area from above.
b. Use the seven-step problem-solving system.
1. Describe the problem as clearly and fully as you can. (It is hard to find one's way around the local school area.)
2. Describe the results you want. (The desired result is a readable map that will assist people to find their way around the area near school.)
3. Gather information.
4. Think of alternative solutions. Determine with the students the best way of producing aerial views (e.g., standing on the roof and doing a technical drawing, climbing a tall tree nearby and sketching, taking a photograph from above using kites, model rockets, helicopters, etc.).
5. Choose the best solution. (Determine the advantages and disadvantages of each method.)
6. Implement the solution you have chosen. Whatever class decides is best and is legal and practical.
7. Evaluate the solution and make necessary changes (Have a person try to follow the map to locate a designated spot.)
c. Use existing maps of a different area as models; use books on drawing and sketching; review aerial photographs; determine how each idea can be used to advantage.
d. Decide, through class consensus, which method to pursue and generate a finished map.

Discuss with the class how this problem differs from the kind of problem a mapmaker would face if he/she wished to map the town in which the school is located. What kind of political problems would be encountered? What types of permits would be needed? Are there any environmental concerns that the mapmaker would have to consider? Could the mapmaker have free access to all parts of the town? Are any of the areas to be mapped dangerous to human health? How often would the map need to be revised? What would necessitate a revision?

An example of a problem needing insight is to determine the next figure in the following series:

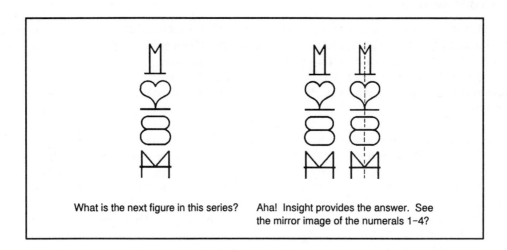

What is the next figure in this series?   Aha! Insight provides the answer. See the mirror image of the numerals 1–4?

To convey the use of insight as a problem-solving technique, refer to a marvelous book titled *Aha! Insight!*, by Marvin Gardener, Scientific American Inc., W.H. Freeman and Company, New York City, N.Y.

While the problem solving project is instructive, the differences between such a problem and "real-world" problems should be pointed out. A classroom problem that involves environmental, social, political, and other concerns would be an excellent vehicle for more advanced classes.

Modeling, a subject that deserves its own book, is touched on in the text as a class of techniques to test solutions to problems without building the complete systems. Depending upon the time available, several of the modeling techniques can be used in the classroom in preparation for solving a problem, or as projects in themselves if a very large system is being simulated.

Chapter 3 is an introduction to system theory at a very basic, non-rigorous level. One of the objectives of the chapter is to provide the student with system thinking as a tool for understanding technology and to help in solving problems. The fundamental concepts of INPUT, PROCESS, OUTPUT, and FEEDBACK are introduced early in the chapter with almost intuitive definitions. A good strategy for reinforcing these concepts would be to suggest different kinds of systems and have the students pick out the different components. Examples might include the operation of tools or kilns in the laboratory, riding a bicycle, and an educational system (in which technological literacy of students is the desired result, and tests, quizzes, and homework provide feedback).

The systems concepts of input, process, output, and feedback apply equally well to many systems other than technological systems. Examples in addition to those given (biological systems and teaching systems) include economic systems, political systems, and social systems. These systems, and many complex technological systems, are easier to understand if they are broken down into their smaller components, or subsystems. It is instructive to use an example of a very complex system, break it down into many subsystems, show how each of the small subsystems is easily understood, and then show how they are combined to form the large system.

A key point in the study of technology (and one that recurs in Chapter 14) is that systems often have multiple outputs. In addition to the intended output (which is often singular), technological systems can produce environmental, political, social, economic, and other outputs: some bad, some good, and some neutral. This concept is key to understanding of the dimensions of technology's impact on society.

Hints for Activities:

1. Students can design and construct a spacecraft that is rocket powered.
2. Create teams to design and build rockets. Each team chooses a name and competes with other teams. Use computer software to fire the rockets and record data. Award prizes for various selected categories.
3. Invite parents to become advisors of rocket teams or companies. Hold a well-publicized launch date with the news media and local officials and parents invited.

Individual Activity:

Assign an out-of-class activity.

    a. Students look in magazines, newspapers, and brochures for pictures and articles relating to spacecraft launches.

    b. Have students assemble a notebook or use these for a bulletin board display. Require students to explain the systems approach in the launch of a rocket. Sketch and label the diagram.

    c. Ask the students to watch the news and read newspapers concerning launches at the time of the assignment. Discuss this with them in class on a selected day.

Cooperative Activity:

Assign teams to research the history of launches of various types of spacecraft:

    a. Teams go to libraries, view videos, and watch news stories.

    b. Each team is to construct a model of a spacecraft of the launch they have chosen.

    c. One person from each team is to present the research and model spacecraft to the class.

    d. Have the class as a group discuss how spacecraft launches entail many systems and subsystems.

Overhead Transparency Masters can be found in Appendix A.

Chapter Review Questions and Answers can be found in Appendix B.

Crosstech Puzzles and Solutions can be found in Appendix C.

Chapter Evaluation Material can be found in Appendix D.

Technology Activity Guidelines can be found in Appendix E.

A General Safety Unit can be found in Appendix F.

THE ELECTRONIC COMPUTER AGE

Chapter 4, Text pages 88 to 119

Chapter Summary

Chapter 4, "The Electronic Computer Age", deals with electronics circuits, integrated circuits, computers as tools, computers as systems, and how computers operate.

Chapter Outline

A. Electronics (p. 89)
B. The Smallest Pieces of Our World (p. 90)
C. Electric Current (p. 91)
D. Electronic Components and Circuits (p. 93)
    1. Printed Circuits
    2. Integrated Circuits
    3. Analog and Digital Circuits
E. Computers (p. 99)
F. What is a Computer System? (p. 101)
    1. The Computer Processor
    2. Memory
    3. Computer Input
    4. Computer Output
G. Computers, Large and Small (p. 107)
H. Computer Software (p. 108)
I. Computers and the System Model (p. 110)
J. Summary (p. 111)
K. Review Questions (p. 115)
L. Crosstech
M. Activities (pp. 116–119)

Major Concepts:

- The use of electronics has completely changed our world in the last hundred years.
- Electric current is the flow of electrons through a conductor.
- Electronic circuits are made up of components. Each component has a specific function in the circuit.
- An integrated circuit is a complete electronic circuit made at one time on a piece of semiconductor material.
- Computers are general-purpose tools of technology.
- Computers use 1s and 0s to represent information.
- Computers have inputs, processors, outputs, and memories.
- Computers operate under a set of instructions, called a program. The program can be changed to make the computer do another job.
- Computers can be used as systems or can be small parts of larger systems.

Student Behavioral Objectives:

Upon completion of the "Computers in Industry" activity, the student should be able to:

1. Use computer software to test machinery, build a factory, design a product, and/or design a car.

2. Explain how use of a more advanced type software is used in industry.

3. Build a model of a product designed on the computer.

4. Display correct use of a computer and computer software.

Hints for Chapter 4

Instructional Suggestions:

a. Announce in advance the activity, materials needed, and outside-of-class assignments.
b. Encourage students to explore types of software relating to designing, testing, and/or modeling products on their personal computers.
c. Order software and videotapes about software for the lab activity.
d. Write up and duplicate instructions as needed for operation of the software.
e. Load and check software.
f. Order materials for the building of CAD design products.

The widespread use of electronics has revolutionized the lifestyle we experience in the industrialized world. This revolution has take place over a relatively short period of time in the history of technology; the transistor was invented in 1947, and modern solid state electronics came into commercial use in the 1950s. If the class is working with a technology time line, particularly one that shows exponential growth, the use of electricity and then electronics can be located on the time line, and will be associated with increases of inventive activity.

One activity that may help students understand the pervasiveness of electronics in our lives is to make a list of ways in which electronic devices are used in everyday applications, and to explain how the same activities were (are) carried out without electronics. A similar activity can be used with electrical appliances, tools, and machines.

The analogy that links electron flow in a wire and water flow through a pipe is meant to help students understand this concept, but its limitations should be recognized when dealing with it in class. The time it takes for an electron entering a wire at the negative pole to return to the positive pole is indeed very short, for the journey occurs almost at the speed of light. This movement of free electrons is much faster than water flow. It should also be noted that the electrons themselves do not move through the wire from end to end at the speed of light; it is the pulse of current created by the chain reaction of electrons flowing from atom to atom that does approach the speed of light.

A very good tie to math and science for academically capable classes is to spend some time with Ohm's Law. Ohms are briefly described on page 92. Practical problems using voltage, resistance, and current can be posed by the teacher and solved quantitatively by the students to increase their familiarity with the relationship among the three variables.

The electronic components described on pages 93-95 are but a few of the component types commonly used to make electronic circuits. These descriptions can be augmented with samples of these and other components, which are commonly available in neighborhood electronics stores, such as a Radio Shack™ distributorship. A simple setup using a battery, inexpensive meters, and various components can be used to demonstrate the properties of some of these components, while others require more sophisticated support equipment. A variety of resistors, some wire, and an appropriately sized current meter could be used to demonstrate and support Ohm's Law calculations; a small light with various resistor values in series with it would give more tangible (but less quantitative) indication of more or less current flow.

If there are no samples of printed circuits or integrated circuits readily available in the lab, it may be possible to obtain some from a local distributor, manufacturer, or business user of electronic equipment. While these may be inoperative or obsolete for practical use, they might serve as excellent physical examples of the way that components are combined to perform tasks. Slides and photographs are available from several sources, including both the Computer Museum in Boston and IBM Corporation, that can be used to show how more and more functions (usually indicated by transistor or gate count) have been fitted into smaller sizes.

The concept of analog vs. digital circuits is one that may be most easily conveyed through the use of examples. Some examples that students might be familiar with are tuners on radios (those with knobs that turn smoothly, activating a pointer, are analog; those that are activated by a keyboard or that have "detent" knobs, often used with a numeric display of frequency, are digital); car speedometers (those that use a pointer that swings in an arc are analog, while those that have numeric readout that can only change in defined increments—usually one mph—are digital).

It is clear that the topic of computers can be given only the most general of treatments in the space of less than one chapter. It is also clear that an understanding of the range of uses of computers is essential to an appreciation of modern technology. The teacher will have to evaluate students' capabilities to grasp the concepts of computing presented in this chapter, and use only portions of the material, or augment it when appropriate. An excellent reference to use in conjunction with this chapter is *Using Computers in an Information Age,* by Brightman and Dimsdale, Delmar Publishers Inc.

At the most basic level, students should understand that computers are tools of modern technology that can be easily programmed (instructed) to provide a particular function. Because the function provided can be changed by changing the program, the computer is a very flexible tool. Computers can be used in all types of technologies to monitor, control, and be part of the process. In some information systems, the computer provides the main functions of the process.

Nearly all modern computers are digital computers. In the early years of computer development, analog computers that solved differential equations and other mathematical problems were used frequently; digital computers are now generally used for these tasks. Analog computers made up of mechanical parts (cams, gears, motors, etc.) controlled the position of guns used aboard ships during the 1940s and 1950s; these have now been replaced by digital electronic computers.

A computer can be thought of as having four components (not including software): a processor (often called a central processing unit (CPU) in large machines, or a microprocessor in computers in which it is a single integrated circuit); memory (which includes main memory—RAM and ROM—and storage—disk, tape, and optical); input (keyboards, tapes, disks, cards, transducers, etc); and output (printers, plotters, terminal screens, etc). Often, input and output are considered together, as they are sometimes interactive or related in some way. This is often referred to as "computer I/O." The depth to which the teacher explores the parts of a computer will depend on the academic ability of the students and the availability of equipment.

Computer software is briefly described on pages 108–110. Computer literacy is not knowledge of programming, but rather an understanding of computer principles and an awareness of the availability and capability of various kinds of programs and programming languages. (There is currently lively discussion on the definition of computer literacy.) Computer software, like the rest of the computer area, has been given only cursory attention in the text; the teacher is encouraged to augment this in class as appropriate.

Hints for Activities:

Individual Activity:

Assign an out-of-class activity.
   a. Students look in magazines, newspapers, and brochures for advertisements and articles about software used for purposes of designing and testing products and managing production systems.
   b. Have students assemble a notebook or use these for a bulletin board display. Require students to group software according to its uses. Include printouts from the software in the notebook.
   c. Ask a number of students to present their findings to the class. Some may bring in software. Have this software demonstrated, so that all students can see what is available.

Cooperative Activity:

Assign teams to use, produce hard copy, and evaluate software relating to designing, testing, and managing of production systems.
   a. Assign each team a type of software and have them try it and do printouts. Require the team to research the availability and types of software related to their assignment.
   b. Set up a competition where each team is to construct a product based on the software they used. Each team is required to submit research on related software, a printout of the product, and the completed product. Have faculty members or individuals from industry judge the competition. Invite the local news media.
   c. Hold a class discussion concerning all findings. Ask the team's spokesperson to relate his or her findings about use of that particular software in industry.

Overhead Transparency Masters can be found in Appendix A.

Chapter Review Questions and Answers can be found in Appendix B.

Crosstech Puzzles and Solutions can be found in Appendix C.

Chapter Evaluation Material can be found in Appendix D.

Technology Activity Guidelines can be found in Appendix E.

A General Safety Unit can be found in Appendix F.

COMMUNICATION SYSTEMS

Chapter 5, Text pages 120 to 135

Chapter Summary

Chapter 5, "Communication Systems", deals with what kinds of communication systems exist and how they work. The purposes, as well as the uses, of the seven technological resources are explained. Graphic and electronic communications are explored.

Chapter Outline

A. What is Communication? (p. 121)
B. Types of Communication (p. 121)
   1. Person-to-Person Communication
   2. Animal Communication
   3. Machine Communication
C. Communication Systems (p. 124)
D. The Process of Communication (p. 125)
E. Designing the Message (p. 125)
F. Resources for Communication Systems (p. 127)
   1. People
   2. Information
   3. Materials
   4. Tools and Machines
   5. Energy
   6. Capital
   7. Time
G. Categories of Communication Systems (p. 130)
H. Summary (p. 132)
I. Review Questions (p. 133)
J. Crosstech
K. Activities (pp. 134–135)

Major Concepts:

■ Communication includes having a message sent, received, and understood.

■ Humans, animals, and machines can all communicate.

■ A communication system has an input, a process, and an output.

■ The communication process has three parts. They are: a transmitter, a channel on which the message travels, and a receiver.

■ Communication systems are used to inform, persuade, educate, and entertain.

■ Communication systems require the use of the seven technological resources.

■ Two kinds of communication systems are graphic and electronic.

Student Behavioral Objectives:

Upon completion of the "International Language" activity, the student should be able to:

1. Explain the impact of an international language on worldwide trade.

2. Give examples and explain the meaning of at least ten signs used in malls and airports.

3. Develop a set of instructions that could be understood by persons speaking any language.

Hints for Chapter 5

Instructional Suggestions:
  a. Announce in advance the activity, materials needed, and outside-of-class assignments.
  b. Ask students to look at signs one week prior to the lab activity.
  c. Invite a guest speaker to address the class about the designing and manufacture of wordless signs.
  d. Prepare and duplicate any information and instructions needed for the activity.
  e. Display posters and copies of signs around the lab.
  f. Order materials for the lab activity.

The section, "Types of Communication," (pages 121–124) is intended to expand students' conceptions of communication to include communication among animals and machines as well as people. As a brainstorming activity or for homework, students may be challenged to list different ways in which people communicate with machines and animals. Students can also be challenged to name examples of machine-to-machine, machine-to-animal, animal-to-animal, and people-to-people communication.

Communications are usually not meaningful unless there is confirmation that the message was received and understood. This can be viewed as the feedback component of a communication system. Exploring examples of this will show that the means of communicating feedback is often quite different from the original means of communication. For example, when a TV or radio broadcast station wants to find out if its message is being received and understood, it may do so through monitors (to see if the signal quality is good), through polls (to see if listeners think that the programming is worthwhile), and by tabulating phone calls and letters from listeners. None of these feedback channels uses TV or radio, as the original path did.

On the other hand, a two-way radio (the type used to communicate with remote mountain ranger stations) uses a radio transmitter at both ends to signal the primary message and the response ("Roger" or "Please repeat the message"). Students can be asked to explain the primary and feedback paths used in different communication systems, including newspapers, people talking on a telephone, machines communicating with computers, etc.

On pages 125 and 126, the general concepts of transmitter, channel, and receiver, used in all communication systems, are introduced. While examples are given for different kinds of communication systems, perhaps the easiest to use as examples to introduce the concepts are telephones, telegraph, speech, and radio/TV communication. In TV communication, the word "channel" is used to refer to a piece (bandwidth) of the electromagnetic spectrum in the air between the transmitter and the receiver.

In explaining what is meant by a communication system, a distinction should be made between the communication process and the entire system. The process component consists of a transmitter, a channel, and a receiver. Several examples are shown on page 126. The entire system includes the specified input, the output, and the feedback loop, as well as the process component.

In introducing students to the uses of communication systems, students might be asked to design a message specifically to inform, persuade, entertain, or educate a target audience. The message may take various forms. For example, one group of students may be assigned to deliver entertainment through an audio medium alone; another group may use only video; another group might be restricted to print media. The effectiveness of each method can be compared by the class.

In defining the difference between graphic and electronic communication, the authors have chosen to focus on the type of message carried by the channel. In graphic communication, the channel carries images or printed symbols. In electronic communication systems, the channel carries electromagnetic signals. Admittedly, trying to make a clear distinction is sometimes difficult. Some systems (e.g., word processing) are underpinned by electronics; however, their final purpose is to generate printed copy, and therefore they have been classified as graphic systems. Some elaboration relative to this distinction by teachers would help students see that things are not always black and white.

Hints for Activities

Individual Activity:

Assign an out-of-class activity.
  a. Students are to spend one week looking for wordless signs. Suggest they look in malls, restaurants, highways, the school, and in their homes. Have them look in magazines, newspapers, and brochures.

b. Have the students assemble a notebook in which they make ten wordless signs they have seen. Then, ask them to create two signs of their own.
c. Ask the students to look for any brochures, labels, etc., for instructions that are wordless. Have them duplicate these in a notebook or portfolio.

Cooperative Activity:

Assign teams to develop a brochure explaining how to use a product. Each team is to have the same product. Set up guidelines for a competition. Explain how in the real world, advertising companies compete in this way for a company's business. Their ability to win contracts determines the profit margin and whether or not they will stay in business.

a. Assign all teams in a class the same product. Each team is to research the product, look for ideas, and come up with thumbnail sketches.
b. In class, have the teams work on development of the brochure. Encourage them to use desktop publishing software to complete the final designs.
c. Ask each team to select a spokesperson who presents the brochure or label to the class.
d. Have people from the advertising industry, art instructors, and/or faculty members judge the brochures or labels.
e. Hold a class discussion concerning all solutions. Discuss what changes could be made in any of the brochures or labels that would enhance or clarify the instructions.

Overhead Transparency Masters can be found in Appendix A.

Chapter Review Questions and Answers can be found in Appendix B.

Crosstech Puzzles and Solutions can be found in Appendix C.

Chapter Evaluation Material can be found in Appendix D.

Technology Activity Guidelines can be found in Appendix E.

A General Safety Unit can be found in Appendix F.

GRAPHIC COMMUNICATION

Chapter 6, Text pages 136 to 165

Chapter Summary

Chapter 6, "Graphic Communication", covers pictorial and orthographic drawings, elements needed for photography, different types of printing processes, word processing, and desktop publishing.

Chapter Outline

A. Introduction (p. 137)
B. Planning and Designing the Message (p. 137)
C. Writing (p. 138)
D. The Invention of Paper (p. 139)
E. Freehand Drawing and Sketching (p. 139)
F. Technical Drawing (p. 140)
G. Computer-Aided Design
H. Photography (p. 144)
I. The Development of Printing (p. 147)
   1. Relief Printing
   2. Gravure Printing
   3. Screen Printing
   4. Offset Printing
J. Word Processing (p. 152)
K. Desktop Publishing (p. 153)
L. Computer Printers (p. 154)
M. Photocopying (p. 156)
N. Summary (p. 157)
O. Review Questions (p. 159)
P. Crosstech
Q. Activities (pp. 160–165)

Major Concepts:

- In a graphic communication system, the channel carries images or printed words.

- Pictorial drawings show an object in three dimensions. Orthographic drawings generally show top, front, and side views of an object.

- The five elements needed for photography are light, film, a camera, chemicals, and a dark area.

- Four types of printing are relief, gravure, screen, and offset.

- Word processing improves office productivity.

- Desktop publishing systems combine words and pictures.

Student Behavioral Objectives:

Upon completion of the "Team Picture" activity, the student should be able to:

1. Explain the process of screen printing.

2. Demonstrate a knowledge of the procedure to follow in screen printing.

3. Design and screen print a product.

4. Discuss with the class the different methods used to form screens and designs for screens.

Hints for Chapter 6

Instructional Suggestions:

    a. Announce in advance the activity, materials needed, and outside-of-class assignments.

    b. Ask students to look at T Shirts or uniform shirts one week prior to the lab activity.

    c. Invite a guest speaker from a graphic communications company to demonstrate screen printing processes.

    d. Prepare and duplicate any information and instructions needed for the activity.

    e. Display screen printed T Shirts, folders, sweatshirts, and hand towels around the lab.

    f. Order materials for the lab activity.

    g. Set up computers with software for the purpose of creating designs.

Printing is a traditional Industrial Arts activity that can be enhanced by making use of many different aspects of technology. This can be demonstrated by handling it as an enterprise: Select a message to be communicated to the rest of the student body; choose an optimum medium (or several media in combination, such as posters or newsletters etc.); select appropriate layout or design for the message; print the message as a group activity; distribute the posters or newsletters; and obtain feedback to find out if the message had the desired effect.

The power of technology in changing the way that people do things is demonstrated in a series of activities developed at the Clarkstown, N.Y. Central Schools. Students are asked to make technical drawings of objects that require straight lines at various angles. Initially, the students are not allowed to use a straightedge of any kind. In order to improve the quality of the drawings, students then design a wood T-square and a plastic triangle, and mass-produce them on a production line, using student-made jigs. The same drawings are then attempted again, and the improvement is demonstrated. CAD stations are then used to create drawings of the same object. Students are then able to see the quality and consistency of CAD drawings, and to experience the quantum steps in capability offered by different technologies. (Thanks to Alan Horowitz of Clarkstown Central Schools.)

Cameras and photography can also offer demonstrations of how modern technology has provided new capabilities. An interesting project is to have students design and build a camera obscura or a pinhole camera. Once students understand the fundamentals of cameras, the addition of a microprocessor to the camera to automatically calculate and control aperture or exposure time can be appreciated.

Desktop publishing is rapidly becoming a popular method of pre-press composition. It is a powerful technology that is within reach of most school budgets. The Apple MacIntosh is the favorite of many desktop publishing enthusiasts, although IBM (MS-DOS) machines with graphics cards can run some DTP packages.

Presently several software programs are available. These include Pagemaker (Aldus Corporation, 411 First Avenue South, Seattle, WA 98104: (206) 662-5500) for MS-DOS and MacIntosh computers, and Ventura (Xerox Corporation) for MS-DOS computers. Others include ReadySetGo (Manhattan Graphics, 63 Varrick Street, New York City, NY 10013; (212) 924-2778, and Ragtime (Orange Micro Inc., 1400 N. Lakeview Avenue, Anaheim, CA 92807; (714) 779-2772) both for the MacIntosh.

Hints for Activities

Individual Activity:

Assign an out-of-class activity.

    a. Students are to spend one week looking at sweatshirts, T Shirts, printed scarfs and towels, and notebooks.

    b. Require the students to turn in five thumbnail sketches of designs for a shirt, towel, or folder.

    c. Have each student select one of his or her 5 sketches and do a final sketch of it on 8 1/2" x 11" paper.

    d. Set up a competition for the best design from each class. The students are to vote on a design and make screens for one or two of the designs.

    e. Each student is to screen print in class a shirt and/or a folder. Use a textile ink for the shirt.

Cooperative Activity:

Form a company with the task of creating a design for a T Shirt that can be mass produced, packaged, and sold to all students for a profit.

a. Have students choose positions in the company or elect positions.
b. Use computer software, if available, to set up an organizational chart. Create a flow chart of the tasks to be performed.
c. Students can have a competition within the company for the best design, which is then given to those assigned to make the screen.
d. Students whose tasks within the company can have dual tasks so that there is a greater amount of efficiency.
e. Tasks for the completed product include managerial, advertisement, design, production, labeling, packaging, quality control, and sales.
f. Upon completion of the project, including the selling, have the entire class conduct an evaluation of the company.

Overhead Transparency Masters can be found in Appendix A.

Chapter Review Questions and Answers can be found in Appendix B.

Crosstech Puzzles and Solutions can be found in Appendix C.

Chapter Evaluation Material can be found in Appendix D.

Technology Activity Guidelines can be found in Appendix E.

A General Safety Unit can be found in Appendix F.

ELECTRONIC COMMUNICATION

Chapter 7, Text pages 166 to 206

Chapter Summary

Chapter 7, "Electronic Communication", deals with communication systems with electronic or electromagnetic signals for a channel. These include the telephone, radio, telecommuting, telegraph, and data communication.

Chapter Outline

A. What is Electronic Communication? (p. 167)
B. Transmitting and Receiving Systems (p. 168)
    1. The Telegraph
    2. The Telephone
    3. Radio
    4. Television
    5. Microwave Communication
    6. Satellite Communication
    7. Fiber Optic Communication
C. Data Communication Systems (p. 180)
    1. Computer Codes
    2. Modems
    3. Data Networks
    4. Facsimile
D. Recording Systems (p. 184)
    1. Phonograph Records
    2. Compact Disks and Video Disks
    3. Tape Recordings
E. Modern Telecommunications Services (p. 194)
    1. Teleconferencing
    2. Telecommuting
    3. Electronic Banking
    4. Two-way Cable TV
F. Summary (p. 191)
G. Review Questions (p. 193)
H. Crosstech
I. Activities (pp. 194–205)

Major Concepts:

- In an electronic communication system, the channel carries an electronic or electromagnetic signal.

- Electricity and electronics have greatly changed communication technology.

- The telephone is the most commonly used form of electronic communication.

- In radio communication, the message is sent through the air from a transmitting antenna to a receiving antenna some distance away.

- Computing and communication technologies are being used together to make powerful new tools for working with information.

- A number of computers or computer devices joined together is called a network.

Student Behavioral Objectives:

Upon completion of the "Beam That Signal" activity, the student should be able to:

1. Explain how lasers are used for communication purposes.

2. Use a laser beam to simulate point-to-point transmission of radio communication systems.

3. Demonstrate safety techniques used when handling a laser.

4. Participate in a class discussion relating to uses of lasers in modern-day communication systems.

5. Calculate degrees of angles from the activity and use the data.

Hints for Chapter 7

Instructional Suggestions:

  a.   Announce in advance the activity, materials needed, and outside-of-class assignments.
  b.   Ask students to watch Discovery channel on television and look in magazines, books, and newspapers for articles relating to uses of the laser in communications.
  c.   Invite a guest speaker from an electronics-related business or industry.
  d.   Prepare and duplicate any information and instructions (especially that concerning safety) needed for the activity.
  e.   Obtain and prepare lab equipment for the activity.

   Electronic communication systems are those that use electricity in some form to store or carry a message. In the modern world, electronics are used in many types of communications systems as well, but that does not make them electronic communication systems. For example, a modern newspaper uses word processors, electronic typesetters, and perhaps satellite communications to distribute information within the organization. The newspaper itself, however, is still made of paper with ink on it, and it is the entity that actually provides communications with the readers. A newspaper, therefore, is a graphic communication system, not an electronic communication system.

   The authors have divided electronic communication systems into two broad categories: transmitting and receiving systems and recording systems. Transmitting and receiving systems immediately send a message over a channel to a receiver. Recording systems store a message in some electronic, optical, or electromagnetic medium for transmission at a later time. Some systems, such as radio and television networks, use a combination of the two, as much of the broadcast programming is previously recorded.

   Telegraph systems make excellent laboratory projects because they combine concepts in electricity, communications, and symbols and codes in hands-on construction and operating activities. More advanced students can be encouraged to progress to radio projects, or even to qualify for amateur radio licenses, if the facilities are available.

   Similarly, telephones are relatively simple devices that may be used in laboratory projects. Modern telephone products often use several technologies in combination. Demonstrations of these in the classroom can help to reinforce the concepts; construction or dismantling of simple telephones and showing how they are connected can also be instructive. It may be pointed out to students that, just as it is now commonplace to have one or more telephones in every home (and, in most offices, one on every desk), it was not always so; similarly, computers or computer terminals are found on most desks in many offices, and are becoming quite common in homes as well.

   The spectrum chart contained in the section on microwave communication shows that the different types of communication often thought of as separate and distinct, namely AM radio, FM radio, TV, shortwave radio, microwave radio, and light communication, are all really part of a continuous spectrum of electromagnetic waves. Satellite communication is simply a specific application of microwave radio.

   One way to use the math and science concepts surrounding wave propagation is to have the students calculate the delay of a radio signal being relayed by a satellite. (Radio signals and light travel at the same speed: 186,000 miles per second. The satellite is approximately 22,500 miles high, so one uplink and one downlink path is 45,000 miles. The one-way delay is thus about 1/4 second. An answer to a question would suffer a two-way delay, or about 1/2 second, quite noticeable in speech.

A discussion about whether satellite communication is better than fiber optic communication for conversations between Europe and New York would then point out the choices made in selecting the right technology for each application. (Note: the delay for the fiber-optic cable is roughly 4,500 miles/ 186,000 miles per second, or about 1/40 second, not noticeable in speech. The satellite delay does not matter at all for TV or other broadcast services; the listener is unaware of it since there is never any response.)

The use of fiber-optic cable for communication may be explained using a simple analogy. Imagine having a light at one end of the fiber-optic cable that can be turned on and off in a code similar to a Morse code. At the far end of the fiber-optic cable, an operator can read the code and decipher the message. Extending this concept, the light source (a LASER or LED) can be turned on and off millions of times per second, and a device that can automatically detect and decode the light pulses is placed at the receiving end. Because speech can be turned into digital pulses (see page 170), it can be sent by this optical on-off code. Because fiber-optic cable can support millions of on-off pulses per second, many telephone conversations can be mixed together (multiplexed) and sent over a single light beam. (Note: a standard digitized telephone conversation is sent at between 32,000 and 64,000 bits per second; typical fiber-optic system signaling rates range from 45 million to almost 500 million bits per second.)

Data communication is a form of machine-to machine communication using digital signaling. Computers use data communication systems to send and receive data to and from printers, remote sensors, data bases, and other computers. Data communication can be quite complex, and the mathematics used to explain it are of a very high level, but the principles can be demonstrated with almost no mathematics, and in very physical terms. "1's" and "0's" can be demonstrated by using lights, sounds, or meters showing voltages. If an oscilloscope is available, it can give a graphic representation of quickly changing electronic voltages. A helpful, relatively low-level text on data communication is *Introduction to Digital Data Communications,* by David H. Stein (Delmar Publishers).

If a modem is available in the laboratory so that communication with electronic mail systems or data bases can be demonstrated, it may be instructive to connect a speaker to the telephone line so that students can listen to the data transmission over the telephone line. Slow-speed modems use one frequency tone to represent "1's" and a different frequency tone to represent "0's". The transmission of data can be imagined very vividly by listening to the tones.

One relatively recent development in the workplace is the providing of communication capability for different kinds of office machines so that they can work together as an integrated system. The integration of these machines with computers (word processors, data base machines, etc.) is called office automation, which is to the information industry what factory automation is to the manufacturing industry.

Recording systems include phonograph records, compact disks, video disks, tape recordings, floppy disks, computer hard disks, and others. In recording systems, information or data is typically converted into optical, magnetic, or mechanical variations in a medium that can later be "read" in a reverse process. It is interesting to note that the newer, high-quality recording systems convert information to digital format before recording. This results in lower noise levels and better preservation of the recorded information. Examples of this are digital tape recordings and compact disks.

Some of the ways in which modern electronic communication technologies are affecting society are described on pages 188 through 192. The examples cited are not a complete list by any means, but are presented as a cross section. Examples pulled from local community practice would serve to bring these concepts into sharper focus if the students are not familiar with the examples described on these pages.

Hints for Activities

1. Add variables to the problem:
   a. Add extra mirrors and relay the signal between them.
   b. Put in a beam splitter and bounce off two different mirrors.
2. Develop a communication system using modeling clay and several 1" mirrored bathroom tiles with the following guidelines and suggestions:
   a. Mirrors cannot be less than one foot from each other.
   b. Draw the pattern out on butcher paper.

    c.   Place clay and mirrors on the pattern.

    d.   Test the system.

3.   Work in teams of two called "Telecommunications Companies."

Individual Activity:

Assign an out-of-class activity.

    a.   Students are to spend one week prior to the lesson searching for information on uses of lasers in communications.

    b.   Require the students to document articles found and their sources. Hand out computer-generated certificates to the students with the best collections.

Cooperative Activity:

Have students choose teams for the lab activity. If only one laser is available, set up other lab activities so that teams can rotate through the activities.

    a.   Each team is to receive a set of instructions and then select a team leader.

    b.   Have each team observe the experiment suggested in the text. One student is to record the information and give that to still another person to report.

    c.   Ask the teams to research how laser communication is used locally and report on this to the class. As a team, they are to compile a report; this is to be done using a word processor.

    d.   Using a modem, communicate with other technology classes or download information from NASA that relates to electronic communication.

Overhead Transparency Masters can be found in Appendix A.

Chapter Review Questions and Answers can be found in Appendix B.

Crosstech Puzzles and Solutions can be found in Appendix C.

Chapter Evaluation Material can be found in Appendix D.

Technology Activity Guidelines can be found in Appendix E.

A General Safety Unit can be found in Appendix F.

PROCESSING MATERIALS

Chapter 8, Text pages 208 to 245

Chapter Summary

Chapter 8, "Processing Materials", covers altering the form of materials to make them more useful. Processing materials involves forming, separating, combining, and conditioning processes. Computers are used to control machines that process materials.

Chapter Outline

A. Material Resources (p. 209)
B. Types of Industrial Materials (p. 210)
    1. Wood
    2. Metal
    3. Ceramics
    4. Plastics
C. Processing Material Resources (p. 215)
D. Forming Processes (p. 215)
    1. Casting
    2. Pressing
    3. Forging
    4. Extruding
    5. Blow Molding and Vacuum Forming
E. Separating Processes (p. 219)
    1. Shearing
    2. Sawing
    3. Drilling
    4. Grinding
    5. Shaping
    6. Turning
    7. Chemical Separating
    8. Filtering
    9. Magnetic Separation
F. Combining Processes (p. 224)
    1. Fastening materials by mechanical means
       a. Nails, Screws, and Rivets
    2. Fastening Materials with Heat
       a. Soldering and Welding
    3. Gluing Materials
    4. Coating Materials
       a. Electroplating
    5. Making Composite Materials
G. Conditioning Processes (p. 228)
    1. Heat-Treating
    2. Mechanical Conditioning
    3. Chemical Conditioning
H. Using Computers to Control Processing of Materials (p. 231)
I. Properties of Materials (p. 232)
    1. Mechanical Properties
    2. Electrical and Magnetic Properties
    3. Thermal Properties
    4. Optical Properties
J. Disposal of Resources (p. 237)
K. Summary (p. 239)
L. Review Questions (p. 241)
M. Crosstech
N. Activities (pp. 242–245)

Major Concepts:

- Processing materials is changing their form to make them more useful to people.
- The technological process brings about the changing of materials from one form to another.
- Technological systems change raw materials into basic industrial materials. Basic industrial materials are then changed to end products.
- Materials are processed by forming, separating, combining, and conditioning.
- Computers can control machines used to process materials.
- Materials are chosen on the basis of their mechanical, electrical, magnetic, thermal, and optical properties.

Student Behavioral Objectives:

Upon completion of the "Checkerboard" activity, the student should be able to:

1. Design and construct a game board.
2. Calculate resources needed to construct a product.
3. Demonstrate the safe use of tools and machines used in separating and combining processes.
4. Apply knowledge and skills to modify an existing product.
5. Participate in an entrepreneurial activity.

Hints for Chapter 8

Instructional Suggestions:

a. Announce in advance the activity, materials needed, and outside-of-class assignments.
b. Ask students to bring in games and to share information regarding these with class members. Information should include type, popularity, consumer target, cost, packaging, and ways to enhance or adapt some of its features.
c. Invite someone from a store that specializes in games; have them present different types of games, and speak about their histories, and popularity.
d. Prepare and duplicate any information and instructions (especially that concerning safety) needed for the activity.
e. Prepare lab equipment for the activity. Make sure all machines are safe to operate with posted safety instructions.

Process can be described in systems terms as that part of the system that combines resources in some ordered way to produce an output. The actual conversion of resources takes place within the technological process. Generally, materials, information, and energy are processed; however, it is also possible to conceive of other resources as being processed (see answer to question one). If we are to be solicitous of fellow humans and the environment, it is important that we monitor the results of the conversion process and assess their short-term and long-term impacts. Processing materials provides a myriad of hands-on activity ideas for students. These range from testing materials (to determine their properties) to producing products. Should teachers wish to engage classes in the construction of products, it is suggested that the project be framed within a problem-solving context. That is, rather than using a step-by-step cookbook approach, permit the student to make decisions about the design, the materials to be used, and the process to be employed. It is suggested that the teacher demonstrate a wide variety of forming, separating, combining, and conditioning processes with the intent of introducing students to the possibilities that exist, instead of trying to make students expert in a few methods. If the student has seen the process and knows it exists, he or she will ask for more specific assistance when ready to use that process. *A broad-based awareness of many processes is preferable to an in-depth study of several.*

Students might suggest why some materials are chosen rather than others, based upon their properties. A materials-testing activity could provide the basis for comparison and selection. Two types of tests can be easily done in a Technology program: indoor and outdoor tests. Indoor tests include destructive and nondestructive mechanical tests which subject a material to stress, those which compare properties like optical clarity and conductivity, and those which involve the analysis of improved properties of alloys and synthetics. Outdoor tests could include coating identical samples of a material with a variety of different finishes, exposing them to the elements for a period of time, and comparing the degree of degradation.

In designing student activities to convey concepts related to processing resources, teachers would want to go beyond processing materials alone. In order to teach how energy is converted from one form to another, students could use the potential energy stored in a spring or rubber band to power a device. Using biomass to generate methane is an example of a bioprocessing activity which can provide a means of waste disposal and create a usable energy source as a byproduct.

The use of computers to control technological processes can be an exciting, up-to-date technology activity. Software such as The Science Tool Kit, available from Broderbund, (17 Paul Drive, San Rafael, CA 94903-2101), processes information received from a variety of sensors. It comes with an interface module that turns the computer into a device that records time, temperature, and light patterns. Among other things, the program can be used to sense motion, analyze seed germination, and test hot and cold plaster. The Danish toy company, Lego Systems, Inc., (555 Taylor Road, Enfield, CT, (800) 243-4870), now produces a set of building block modules that can be combined to create vehicles, plotters, and robots that are computer controlled with an inexpensive interface. With the use of the Kelp card, Apple computers can be easily interfaced to real-world devices like the Robotix toy.

Hints for Activities

1. Design different types of game boards.
2. Design and construct, using a variety of materials and processes, products such as a logo, crest, or flag for the technology organization, school organizations, and/or local groups.
3. For an entrepreneurial activity, create, mass produce, package and sell games.

Individual Activity:

Assign an out-of-class activity.
   a. Students are to spend one week prior to the lesson looking at games in stores, trying out games they may have at home, and talking to friends about games.
   b. Students are to choose one game and document information such as the materials used for production, the type of packaging, the cost, the game's reputation, and its adaptability to modifications.
   c. Each student is to briefly address the class concerning his or her findings. Those who have chosen the same game are to contribute their findings at the same time.
   d. For extra credit, you could at this point ask students to design and create a game and later present it to the class for evaluation.

Cooperative Activity:

Use this activity as an entrepreneurial activity:
   a. Have students select a name for their company.
   b. As a class, discuss what types of processes will be used and what types of employees are needed. Hold an election and/or select positions.
   c. Determine as a class whether or not you will be marketing the checkerboards and selling stock.
   d. Complete an organizational chart and a flow chart of the company's operations.
   e. Complete a test run of one of the products. Evaluate the overall operation and modify it for efficiency.
   f. Produce, package, and market the product.
   g. Evaluate the entrepreneurial activity. Discuss profit/loss margins, safety, quality control, and how this relates to the real world.

Overhead Transparency Masters can be found in Appendix A.

Chapter Review Questions and Answers can be found in Appendix B.

Crosstech Puzzles and Solutions can be found in Appendix C.

Chapter Evaluation Material can be found in Appendix D.

Technology Activity Guidelines can be found in Appendix E.

A General Safety Unit can be found in Appendix F.

MANUFACTURING

Chapter 9, Text pages 246 to 275

Chapter Summary

Chapter 9, "Manufacturing", deals with mass production, the use of the seven resources in manufacturing, automation, and the impact of robots and computers on the manufacturing industry.

Chapter Outline

A. Production Systems (p. 247)
B. Manufacturing Systems (p. 248)
   1. The Craft Approach
   2. The Factory System
C. Mass Production and the Assembly Line (p. 250)
   1. Jigs and Fixtures
D. Impacts of the Factory System (p. 252)
E. The Business Side of Manufacturing (p. 253)
F. The Entrepreneurs (p. 254)
G. Resources for Manufacturing Systems (p. 254)
   1. People
   2. Information
   3. Materials
   4. Tools and Machines
   5. Energy
   6. Capital
   7. Time
H. How Manufacturing is Done (p. 259)
I. Ensuring Quality in Manufacturing (p. 260)
J. Automated Manufacturing (p. 261)
   1. Robotics
   2. CAD/CAM
   3. Computer-Integrated Manufacturing (CIM)
   4. Flexible Manufacturing
   5. Just-in-Time Manufacturing
K. Summary (p. 268)
L. Review Questions (p. 269)
M. Crosstech
N. Activities (pp. 270–275)

Major Concepts:

- Production technologies fill many of people's needs and wants by means of manufacturing and construction systems.
- Manufacturing is making goods in a workshop or factory. Construction is building a structure on a site.
- Mass production and the factory system brought prices down. People were able to improve their standard of living.
- Manufacturing systems make use of the seven types of technological resources.
- There are two subsystems within the manufacturing system: the material processing system and the business and management system.
- Automation has greatly increased productivity in manufacturing.
- Computers and robots have improved product quality while bringing manufacturing costs down.

Student Behavioral Objectives:

Upon completion of the "An Entrepreneurial Company" activity, the student should be able to:

1. Explain the meaning of the word "entrepreneur."
2. Complete a flow chart of operations and an organizational chart.

3. Explain the meaning of profit/loss margin.
4. Participate as a member of a team to produce a profit.
5. Discuss the importance of working as a team to increase productivity.
6. Practice safe use of tools and equipment
7. Use word processing software for purposes of developing labels, instruction manuals, and bar codes.

Hints for Chapter 9

Instructional Suggestions:

a. Announce in advance the activity, materials needed, and outside-of-class assignments.
b. Ask students to visit stores and look at locker organizers and other products. Study the products, as well as the product's materials, labeling, and packaging.
c. Invite an individual from a store or manufacturing industry to address manufacturing of products, organization of the company, quality control, packaging, and marketing.
d. Prepare and duplicate any information and instructions (especially that concerning safety) needed for the activity.
e. Prepare lab equipment for the activity. Make sure all machines are safe to operate with posted safety.

Teachers may wish to read an excellent and thought-provoking essay on industrialization that appears in H.G. Wells' *Short History of the World,* published by Penguin Books in 1936. This essay asserts that the Industrial Revolution was more a social process (resulting in a changed human condition) than it was a mechanical process of change (resulting in assembly-line techniques). Wells points to the factory method which was extant even during the days of Augustus in Rome. The major impact of the industrial revolution, he feels, was the transition of humans from a "drudge" to an intelligent resource. Toffler holds a similar view. He believes that the American Civil War was caused by the collision between the declining agricultural society and the emerging industrial society, rather than between North and South factions, or between freedom and slavery.

Some very major elements of the factory system include the division of labor, substitution of machine power for human power, and the standardization of parts. Factories brought about synchronization. Time began to become a much more important resource as factories could employ labor in shifts and operate all day and all night. The work day and the work week were defined by the factory system. During the agricultural era, the work period was determined by the solar cycle.

One important point that teachers might wish to stress is the lack of transferability of the school laboratory production setting to the real world. Students should be helped to understand that economics plays more of a role than meets the eye in a school production enterprise. In the real world, producers must pay real wages, incur insurance and utility costs, and compete with foreign producers. The student should be helped to understand that labor costs in the United States might make many production enterprises nonprofitable unless companies set up shop offshore and employ foreign workers. It may be startling for students to learn how low typical labor rates are in some third-world nations. Per capita income can be found in *The World Almanac* or in a good encyclopedia. A student-generated chart comparing per capita income worldwide would provide food for good classroom discussion.

A project might be undertaken to determine where most common items are produced today, which are produced in the United States, and how costs compare with those produced abroad. Students might be asked to discuss the advisability of setting up a local company to produce craft items, electronic consumer goods, or household goods.

Production activities can involve a wide variety of items, not limited to those made from traditional materials. Foodstuffs can be grown and/or processed by cooking, drying, pickling, and canning. Items of clothing can be manufactured and screen printed; writing tablets, calendars, and other printed materials can be produced; cosmetics, perfumes, skin creams, and flavored cough syrups can be manufactured safely in the school lab. All of these products can be attractively packaged, and sold for a profit.

Teachers may wish to involve students in a youth leadership activity. Producing items for charitable purposes is an activity well within the scope of a Technology program. For example, toys for less fortunate children are always appreciated during holiday periods.

A Tabletop Technology robotic work cell kit can be purchased From Creative Learning Systems, Inc., 9889 Hibert, Suite E, San Diego, CA 92131 (1-800-621-0852). This work cell enables construction of a simulated self-contained manufacturing unit that contains a tabletop robot and models of other automatic, integrated manufacturing devices that simulate material handling, machining, and automated spraying. The price is around $3500.

Computer software that would assist teachers to convey manufacturing concepts include the following:

■ *The Factory* is a program that permits users to model the manufacture of a part using various machines. Available from: Sunburst Communications, 39 Washington Avenue, Pleasantville, NY 10570 (914-769-5030, 800-431-1934).

■ *Car Builder* is a program that involves aerodynamically modeling a car and then using the computer generated print-out as a template to construct the vehicle. Available from: Sunburst Communications (see above).

■ *Chevy Tech* is a program that allows the user to price the cost of a new Chevrolet with various options. The disk is available free from: Chevy Tech, POB 2054, Warren, Michigan 48090-2054.

■ *The Toy Shop* permits the user to design mechanical models of an antique truck, a carousel, a glider, a catapult, a steam engine, a crane, and a host of other devices. Available from: Broderbund Software, 17 Paul Drive, San Rafael, CA 94903.

Hints for Activities

1. Choose an activity other than the locker organizer, such as benches for the school courtyard or spirit signs.
2. Run time studies on the time required to complete one of the products. Then, have each student complete the product on an individual basis. Compare the time and costs. Are there any differences?
3. Videotape the entire process and have the class view the videotape and determine a more efficient lab setup and use of employees.
4. Include a marketing department. Label, package, advertise, and sell the products for a profit.

Individual Activity:

Assign an out-of-class activity.

a. Students are to spend one week prior to the lesson visiting stores, studying products, and taking mental notes on the materials used, labeling, and packaging of products. They will then share what they have learned in a class discussion.
b. Students are to choose one product and evaluate it for its durability, labeling, packaging, and usefulness.
c. Each student is to briefly address the class concerning his or her findings in relationship to how this information can be applied to the entrepreneurial activity.
d. If any of the students have locker organizers, ask them to bring these to class for the class to study.

Cooperative Activity:

Use this activity as an entrepreneurial activity:

a. Have students select a name for their company.
b. As a class, discuss what types of processes will be used and what types of employees are needed. Hold an election and/or select positions.
c. Determine as a class whether or not you will be marketing the product and selling stock.
d. Complete an organizational chart and a flow chart of the company's operations.
e. Complete a test run of one of the products. Evaluate the overall operation and modify it for efficiency.
f. Produce, package, and market the product.
g. Evaluate the entrepreneurial activity. Discuss profit/loss margins, safety, quality control, and how this relates to the real world.

Overhead Transparency Masters can be found in Appendix A.
Chapter Review Questions and Answers can be found in Appendix B.
Crosstech Puzzles and Solutions can be found in Appendix C.
Chapter Evaluation Material can be found in Appendix D.
Technology Activity Guidelines can be found in Appendix E.
A General Safety Unit can be found in Appendix F.

CONSTRUCTION

Chapter 10, Text pages 276 to 301

Chapter Summary

Chapter 10, "Construction", presents construction as a system with subsystems combining resources to provide a structure as the output. The chapter deals with construction sites, foundations, superstructures, and structures such as bridges, buildings, dams, harbors, roads, towers, and tunnels.

Chapter Outline

A. Construction Systems (p. 277)
B. Resources for Construction Systems (p. 278)
   1. People
   2. Information
   3. Materials
   4. Tools and Machines
   5. Energy
   6. Capital
   7. Time
C. Selecting the Construction Site (p. 284)
D. Preparing the Construction Site (p. 285)
E. Building the Foundation (p. 285)
F. Building the Superstructure (p. 286)
G. Types of Structures (p. 287)
   1. Bridges
   2. Buildings
   3. Tunnels
   4. Roads
   5. Other structures (airports, canals, dams, harbors, pipelines, and towers)
H. Renovation (p. 292)
I. Summary (p. 294)
J. Review Questions (p. 295)
K. Crosstech
L. Activities (pp. 296–301)

Major Concepts:

- Construction refers to producing a structure on a site.
- A construction system combines resources to provide a structure as an output.
- Three subsystems within the construction system are designing, managing, and building.
- Construction sites must be chosen to fit in with the needs of people and the environment.
- A foundation is built to support a structure.
- The usable part of a structure is called the superstructure.
- Structures include bridges, buildings, dams, harbors, roads, towers, and tunnels.

Student Behavioral Objectives:

Upon completion of the "Dome Construction" activity, the student should be able to:

1. Explain the difference in construction between domed and other types of structures.

2. Tell how and where dome structures are used.

3. Justify the use of a dome type structure as a dwelling.

4. Explain which geometric shapes are used, and how they are used, in the construction of a dome.
5. Using a variety of materials, construct a dome.
6. Practice the safe use of tools and equipment in the lab.

Hints for Chapter 10

Instructional Suggestions:

a. Announce in advance the activity, materials needed, and outside-of-class assignments.
b. Ask students to look in magazines and newspapers for pictures of different types of structures, particularly domes. Have them bring these to class.
c. Prepare and duplicate any information and instructions (especially that concerning safety) needed for the activity.
d. Invite an architect to address the class concerning domed structures. Videotape the talk.
e. Display models of domes and other structures.
f. Prepare lab equipment for the activity. Make sure all machines are safe to operate with posted safety instructions.

Competitive problem-solving events stimulate the creative juices and can be used to advantage in construction-related activities. A competitive tower-building event related to question 9 in the text would provide students with several pounds of newspaper, a roll of masking tape, and some basic techniques of tower construction (such as the use of struts). Break the class into teams and give them a time limit to see who can build the tallest self-supporting tower.

Bridge-building competitions are always exciting for students. Students may be provided with a basic kit of wooden struts and glue, and challenged to construct a bridge that spans a given distance and can support more weight than bridges built by classmates. Some mathematical and scientific insight should be brought to bear on the solution. Students might be armed with knowledge of triangulation, strength of materials, and use of adhesives.

Teachers may wish to engage in a class project that would culminate in a structure for the school grounds. Ideas could include a bus shelter, a concrete and wood park bench, a bicycle rack embedded in a concrete foundation, and a gazebo. The area around the project could be landscaped. A popular idea is the construction of a modular tool shed which could be raffled off or sold at auction to community members.

The DaVinci Graphic Series is sold by the Hayden Software Company, 600 Suffolk Street, Lowell, MA 01853. It consists of a series of MacIntosh MacPaint documents containing images that have been redrawn by skilled artists. Each image can be rescaled, stretched, flipped, or inverted. This set is good for floor plans, exteriors, and landscaping.

Abracadata, Ltd., (POB 2352, Eugene, OR 97492) produces a three-part software package. Components include architectural design, interior design, and landscape design programs. Each program costs $70.

Hints for Activities

1. Choose to build a full-sized dome which can serve as part of playground equipment or be used as a greenhouse.
2. Construct a dome and install a hydroponic system to be used to grow plants. Conduct studies in hydroponics.
3. Conduct studies of domes. Compare construction costs of related structures, such as athletic stadiums (example: compare the Tacoma Dome in Tacoma, Washington, with the Astrodome in Houston, Texas). How were decisions reached to justify the choice of this type of structure? What are the advantages and disadvantages?
4. Have students construct the dome on an individual basis and then construct it as a cooperative effort. Compare time required and efficiency in use of materials. Was there a difference?

Individual Activity:

Assign an out-of-class activity.

a. Students are to spend one week prior to the lesson looking in magazines and newspapers for articles and pictures about domes.

b. Ask some of the students to briefly address the class concerning their experiences with or knowledge of dome-type structures.

d. For extra credit, have the students construct a dome outside of class using a variety of materials, experimenting with different sizes of triangles.

e. If students know of an existing dome, have them visit the dome if possible; better yet, take the entire class for a visit. If this is not possible, videotape a tour through the dome.

Cooperative Activity:

Use this activity as a class or team project:

a. Form building teams. Each team is to choose the materials and size of the dome they wish to construct.

b. Each team is to select a name for their company and elect the roles they wish to play.

c. Organize a competition where teams will compete for prizes such as largest dome, most creative dome, dome built with the most recycled materials, dome with the wildest decor, and the most structurally sound dome.

d. Have each team figure costs, time, and amounts of materials needed. This is to be put on display along with their plans. This will be a part of a final report.

e. Display the domes in the school, at the chamber of commerce, at building shows, or wherever possible. Get the news media to cover the event.

Overhead Transparency Masters can be found in Appendix A.

Chapter Review Questions and Answers can be found in Appendix B.

Crosstech Puzzles and Solutions can be found in Appendix C.

Chapter Evaluation Material can be found in Appendix D.

Technology Activity Guidelines can be found in Appendix E.

A General Safety Unit can be found in Appendix F.

BUILDING A STRUCTURE

Chapter 11, Text pages 302 to 341

Chapter Summary

Chapter 11, "Building a Structure", deals with the actual construction of the building and the progression of the parts: footing, foundations, floors, walls, roof, utilities, insulation, and then finishing. Manufactured housing and the effects of wind are also covered.

Chapter Outline

A. House Building (p. 303)
B. Preconstruction (p. 304)
   1. Picking the Location
   2. Building Permits and Codes
C. Constructing the Footing and Foundation (p. 305)
   1. The Footing
   2. The Foundation Wall
D. Building the Floor (p. 309)
E. Framing the Walls (p. 310)
F. Sheathing (p. 312)
G. Roof Construction (p. 313)
   1. Framing the Roof
   2. Roof Trusses
   3. Roofing
      a. Asphalt Shingles
      b. Wooden Shingles and Shakes
H. Insulating the House (p. 318)
I. Finishing the House (p. 319)
   1. Exterior Finishing
   2. Interior Finishing
   3. Paint
J. Installing Utilities (p. 321)
K. Plumbing (p. 322)
L. Heating Systems (p. 323)
M. Electrical Systems (p. 324)
   1. Series and Parallel Circuits
   2. Grounding Electrical Equipment
N. Manufactured Housing (p. 327)
O. Commercial Structures (p. 329)
   1. The Effect of Wind on Tall Buildings
   2. Stiffening the Building
P. Summary (p. 331)
Q. Review Questions (p. 333)
R. Crosstech
S. Activities (pp. 334–341)

Major Concepts:

■ Buildings are constructed in steps. The footing, foundation, floors, walls, and roof are built. Then the utilities and insulation are installed. Finally, the structure is finished.

■ The footing is the base of the foundation. It spreads the weight of the structure over a wider area of ground.

■ The foundation walls support the whole weight of the structure and transmit it to the footing.

■ Walls transmit the load from above to the foundation. They also serve as partitions between rooms.

- The roof protects the house against the weather and prevents heat loss.
- Insulation helps keep the temperature of the house constant.
- Utilities include plumbing, electrical, and heating systems.
- Manufactured houses are built in a factory and taken to the construction site.
- Wind effects must be taken into account in designing and building a skyscraper.

Student Behavioral Objectives:

Upon completion of the "Prefab Playhouse" activity, the student should be able to:

1. Explain the difference between a house built completely on site and one that is prefabricated.
2. Use software and/or be able to estimate the amounts of materials needed to construct a model playhouse.
3. Design, estimate, plan, and construct a model playhouse.
4. Practice safe use of hand and machine tools.
5. Discuss in class the importance of housing as a budgeted item.
6. Explain the basic parts of a mortgage.

Hints for Chapter 11

Instructional Suggestions:

    a. Announce in advance the activity, materials needed, and outside-of-class assignments.
    b. Ask students to look in magazines and newspapers for pictures of different types of playhouses. Ask students to visit stores or look in catalogues for playhouses.
    c. Prepare and duplicate any information and instructions (especially that concerning safety) needed for the activity.
    d. Invite a model hobbyist to come and talk about making models. You could also ask someone from a toy store to bring and show to the class a number of models.
    e. Display models of playhouses.
    f. Prepare lab equipment for the activity. Make sure all machines are safe to operate with posted safety instructions.

    The Center for the House (Executive Director: Paul F. Kando); 2232 Decauter Place, NW, Washington, DC 20008) is a good source of information about manufactured housing in Europe and Japan. For information about Swedish factory-built housing, contact Svenska Hus, USA Inc., PO Box 1150, Marshfield, MA 02050; 617-837-9044 (Frank Nee). The Forest Products Laboratory of the United States Department of Agriculture can provide information about new construction materials and techniques. Brochures and photos about the Truss Framed System are available from them (Forest Product Laboratory, USDA Forest Service, PO Box 5130, Madison, WI 53705). For information about careers in the construction industry, contact the Associated General Contractors of America (1957 E. Street, N.W., Washington, DC 20006).

Hints for Activities

1. Design and construct a full-sized playhouse.
2. Design and construct a small booth to be used at sporting events, school fairs, or in public display areas.
3. Form a construction company; design, construct a prototype of, and get orders for model playhouses. Construct and sell the houses, using a student-constructed contract with the buyers.
4. Design different outside finishes for the same product as in flexible manufacturing.

Individual Activity:

Assign an out-of-class activity.
    a. Students are to spend one week prior to the lesson looking in magazines, catalogues, and newspapers for articles and pictures of model playhouses.

b. Ask some of the students to briefly address the class concerning their findings about model playhouses.

d. For extra credit, have the students construct a model playhouse outside of the class using a variety of materials.

Cooperative Activity:

Use this activity as a class or team project:

a. Form a construction company.
b. Have students run for positions and/or choose positions in the construction company.
c. Have students submit designs and then choose the model they wish to construct.
d. The estimator is to work with the architect and (with the help of the class and guidance from the teacher) finalize a list of materials and tools and machines required.

5. Construct and market the playhouse. Display the playhouse and take orders for more.
6. Get the news media to cover the project. Donate one of the playhouses to an orphanage or child care center.

Overhead Transparency Masters can be found in Appendix A.

Chapter Review Questions and Answers can be found in Appendix B.

Crosstech Puzzles and Solutions can be found in Appendix C.

Chapter Evaluation Material can be found in Appendix D.

Technology Activity Guidelines can be found in Appendix E.

A General Safety Unit can be found in Appendix F.

MANAGING PRODUCTION SYSTEMS

Chapter 12, Text pages 342 to 370

Chapter Summary

Chapter 12, "Managing Production Systems", covers information relating to the management of production systems. This includes the tasks of engineers, architects, general contractors, a marketing department, a financial department, and a service department.

Chapter Outline

Major Concepts:

■ A company's management must coordinate the work of different departments to make the best use of resources.

■ Architects design a building's shape and choose materials for it.

■ Engineers design a building's structure and its major systems.

■ A general contractor directs the work of many different people on a building project.

■ A company's marketing department decides what market the company will make products for, what features those products will have, and how they are to be sold.

■ A company that makes a product should provide service and technical information about the product after it has been sold.

■ Whether a company survives and does well depends on its financial management.

■ Managing includes planning, organizing, leading, and controlling.

Student Behavioral Objectives:

Upon completion of the "One, if by Land. . ." activity, the student should be able to:

1. Participate as a member of a manufacturing company.

2. Explain the function of management, labor, and marketing departments.

3. Use measuring tools to accurately measure parts needed for the product.

4. Use metalworking tools and machines.

5. Follow safety rules in all lab procedures.

6. Explain how mass production differs from the craftsperson techniques.

Hints for Chapter 12

Instructional Suggestions:

a. Announce in advance the activity, materials needed, and outside-of-class assignments.
b. Ask students to look in magazines and newspapers for pictures of lamps. Ask students to visit stores or look in catalogues for pictures of lamps.
c. Prepare and duplicate any information and instructions (especially that concerning safety) needed for the activity.
d. Invite someone from management in the building industry to talk to the class concerning PERT and Gantt Charts.
e. Display models and pictures of lamps.
f. Prepare lab equipment for the activity. Make sure all machines are safe to operate with posted safety instructions.

The role of a manager in a production system (or any other system) is one which is perhaps least understood by students. They may have a vague knowledge that managers somehow control what goes on, but since a manager's work does not directly and instantly produce physical products (as does an assembly line worker's or a construction worker's), the notion takes more imagination to grasp.

This chapter deals with the manager's role in the traditional way of breaking down into planning, organizing, motivating, and controlling, on an introductory level. (An excellent text that goes into specifics on management in this way is *Management*, by J.A. Stoner and C. Wankel, Prentice-Hall, 1986.) The chapter describes specific management functions in production systems. It should be pointed out throughout the chapter that many of the management jobs and functions in manufacturing systems are very similar in concept to those in construction systems, though specific tasks are somewhat different and (for simplicity) are treated separately in the chapter.

Part of the intent of the chapter is to help students become more familiar with the duties and requirements of different job categories. The students may have heard terms such as architects, engineers, contractors, marketing managers, etc., but may not have a clear understanding of their job functions. This chapter provides an opportunity to integrate a great deal of career awareness into the course.

On pages 343 and 344, engineers are discussed relative to their role in designing buildings. While it is not covered in depth in the text, it may be pointed out to the students that manufacturing systems also use electrical, mechanical, and other engineers to design products, and industrial, manufacturing, quality, and other engineers to design the assembly processes to build them. While the general education of an electrical engineer who works as an electronic product designer may be very similar to (or the same as) one who designs power systems for buildings, their after-college, on-the-job training and further education will be quite different.

Career information can be obtained from professional societies serving each of the areas mentioned. Several engineering societies, including the Institute of Electrical and Electronic Engineers, the American Society of Mechanical Engineers, the American Society of Civil Engineers, the American Society of Chemical Engineers, and others have headquarters in the United Engineering Building, 345 E. 47th Street, New York, NY 10017. The National Society of Professional Engineers is headquartered at 2029 K Street, N.W., Washington, DC 20006.

Other professional societies that may be helpful include:

■ The American Institute of Architects, 1735 New York Avenue N.W., Washington, DC 20006.

■ The Association of General Contractors of America, 1957 E Street N.W., Washington, DC 20006.

■ The American Management Association, 135 West 50th Street, New York, NY 10020.

The importance of scheduling and project tracking cannot be overemphasized on very large projects that involve tens or hundreds of millions of dollars, thousands of workers, and several years of time. Citing examples of such construction or manufacturing projects that happen to be in the local news or that may be nearby is an excellent way to make these problems seem more relevant to the students. Two of the Review Questions ask students to use the most popular scheduling/tracking tools, the PERT chart and the Gantt chart (named after Henry Gantt, an early industrial engineer), to model class activities. This helps the students to develop a goal orientation at the same time as learning these tools.

Once a construction project is completed, the management of the building and the site changes in nature, but management must continue. Maintaining the physical integrity and appearance of the building, as well as its major systems (heat, power, air conditioning, plumbing, etc.) is a full-time job for many people in large buildings. Evidence of this activity can be pointed out in the certificates of inspection for elevators, fire extinguishers, and so on.

Manufacturing provides a good background for an activity involving an assembly line. If enough students are available, one or more can provide a quality control function by carefully measuring size, weight, hole position, hole size, or some other feature of parts that are made, and plotting the results on graph paper. Using this plotted data, the basics of probability distribution can be qualitatively (if not quantitatively) described.

If the assembly line activity is used in the class, a preliminary task can include the preparation of a business plan, including a cash flow analysis similar to the one on page 354. If a computer with a spread sheet program is available, the cash flow analysis can be prepared on the spreadsheet. When the project is done, the actual results can be compared to the projected costs, and it can be pointed out that this feedback makes the management of the company a closed-loop system.

A discussion of OSHA and EPA rules can include the fact that they protect the safety of workers and the environment, as well as the fact that adhering to them and administering them makes products cost more. Thus, consumers pay for this government-mandated protection.

Hints for Activities

1. Require each student to design and construct a lamp with a variation on the one that was mass produced.
2. Develop Gantt and PERT charts to show the scheduling and monitoring of the production of the lamp.
3. Establish a marketing department that advertises and takes orders from classmates and teachers. Package and sell the product.
4. Create an alternate design and use a variation of materials to build a different type lamp.

Individual Activity:

Assign an out-of-class activity.

    a.   Students are to spend one week prior to the lesson looking in magazines, catalogues, and stores for pictures of lamps.

    b.   Ask some of the students to briefly address the class concerning their findings about lamps, i.e. materials, shapes, and types.

    d.   For extra credit, have the students bring in discarded lamps to be used as examples of designs and materials used in making lamps.

Cooperative Activity:

Use this activity as a class or team project:

    a.   Form a manufacturing company. Have the class run for and select various positions and then flow chart the processes to be followed. Also have the class develop both a Gantt and PERT chart.

    b.   Have students run for positions and/or choose positions in the manufacturing company.

    c.   Have students submit lamp designs and choose the model they wish to construct.

    d.   Develop a list of tools and materials needed and obtain these.

    e.   Have the company manufacture a prototype, examine, and modify it if necessary.

    f.   Manufacture, package, and market the lamps.

    g.   Discuss profit and loss margin and modifications that could enhance the lamp. Finally, conduct an evaluation of the entire production effort.

Overhead Transparency Masters can be found in Appendix A.
Chapter Review Questions and Answers can be found in Appendix B.
Crosstech Puzzles and Solutions can be found in Appendix C.
Chapter Evaluation Material can be found in Appendix D.
Technology Activity Guidelines can be found in Appendix E.
A General Safety Unit can be found in Appendix F.

ENERGY

Chapter 13, Text pages 372 to 399

Chapter Summary

Chapter 13, "Energy", covers general information about energy. It includes potential and kinetic energy, limited and unlimited energy sources, and renewable and nonrenewable energy sources.

Chapter Outline

A. Introduction (p. 373)
B. Work and Energy (p. 373)
   1. Work
   2. Energy
   3. Energy in Our Modern Society
C. Limited Energy Resources (p. 377)
   1. Fossil Fuels
   2. Nuclear Fuels
D. Unlimited Energy Resources (p. 383)
   1. Solar Energy
   2. Wind Energy
   3. Gravitational Energy
   4. Geothermal Energy
   5. Fusion
E. Renewable Energy Resources (p. 387)
   1. Human and Animal Muscle Power
   2. Biomass
   3. Wood
F. Summary (p. 389)
G. Review Questions (p. 391)
H. Crosstech
I. Activities (pp. 392–399)

Major Concepts:

- Work done on an object is equal to the distance it moves multiplied by the force used in the direction of the motion.
- Energy is the ability to do work. It is the source of the force that is needed to do work.
- Kinetic energy is the energy of a moving object. Potential energy is the energy an object has because of its position, shape, or other feature.
- Potential energy can be changed into kinetic energy and kinetic energy can be changed into potential energy.
- The principle of conservation of energy states that energy cannot be created or destroyed, but it can be changed from one form to another.
- Energy sources are limited, unlimited, or renewable.
- Most of the energy used in the United States today comes from limited energy sources.

Student Behavioral Objectives:

Upon completion of the "Hot Dog!" activity, the student should be able to:

1. Explain what a parabolic reflector is and how it works.

2. Determine the focal point of a parabolic reflector.

3. List ways parabolic reflectors can be used.

4. Use tools and machines to construct a parabolic reflector.

5. Follow all safety rules and procedures in the lab.

6. Discuss ways industry uses parabolic reflectors.

Hints for Chapter 13

Instructional Suggestions:

a. Announce in advance the activity, materials needed, and outside-of-class assignments.

b. Ask students to look in magazines, newspapers, and television for information concerning solar cooking. Have them contact power companies for information.

c. Prepare and duplicate any information and instructions (especially that concerning safety when using solar cookers) needed for the activity.

d. Invite a guest speaker and judges from the utility company for the solar cookout.

e. Display models and articles about parabolic cookers.

f. Prepare lab equipment for the activity. Make sure all machines are safe to operate with posted safety instructions.

The relationship among work, energy, and power is explained at the beginning of this chapter and the next. Work and energy are described in Chapter 13, and power is described in Chapter 14. To have done work, a force must be exerted to move an object over a distance. In order to give students an idea of the order of magnitude of work measurements, the students can push on a scale to indicate how much force they generate as they push a heavy box across the floor, and work can be calculated. This activity can be timed to calculate power (horsepower) if done in conjunction with the next chapter.

The conversion of energy from one form to another is one of the most important jobs of technology. The principle of conservation of energy, and the notions of potential and kinetic energy, are science concepts that are central to dealing with energy in a technology lab.

The kinetic energy calculations shown in the picture at the bottom of page 374 can be recreated using the equation:

$$\text{Energy} = 1/2 \, (\text{Mass}) \times (\text{Velocity})^2 \, (\text{squared})$$

Energy will be in foot-pounds if mass is in pounds and velocity is in feet per second. Mass is the weight of an object divided by the gravitational constant, 32.2 feet per second per second (feet per second squared).

Studying the natural resources of energy can lead to discussions and projects in energy conservation and recycling. It is an important observation that most of the energy used in this country comes from limited resources (pp. 376, 377). The energy that comes from renewable sources in the pie chart on page 376 includes energy from unlimited sources, a very small fraction of the total. Unlimited energy sources (solar, wind, etc.) are often put to good use in technology lab projects, such as the solar hot dog cooker activity in this section.

The energy industries in general have excellent books, movies, tapes, and other educational material available at little or no cost to technology teachers. A good source is the power company serving your area; many have catalogs listing the materials available. Other sources are:

■ American Gas Association, 1515 Wilson Blvd. Arlington, VA 22209

■ American Nuclear Society, 555 North Kensington Ave. LaGrange Park, IL 60525

■ American Petroleum Institute, 1220 L Street, N.W., Washington, DC 20006

■ Atomic Industrial Forum, 7101 Wisconsin Ave. Bethesda, MD 20814

■ Center for Renewable Resources, Suite 638, 1001 Connecticut Avenue N.W., Washington, DC 20036

■ Edison Electric Institute, Education Service Department, 1111 19th Street N.W., Washington, DC 20036

■ Electric Power Research Institute, P.O. Box 10412, Palo Alto, CA 94303

■ United States Department of Energy, Office of Public Affairs, Washington, DC 20585

■ Wood Energy Institute, Box 800, Camden, ME 04843

Hints for Activities

1. Have students vary the size and type of solar cooker.
2. Work with science classes as teams to develop the fastest cooking solar cooker. Hold a contest, giving each team the same type and amount of food to cook. Judge the results for time and quality of cooking.
3. Work with math classes to figure the dimensions needed to construct an effective solar cooker.
4. Develop a videotape explaining the procedure so that someone else could follow the procedures to build their own solar cooker.

Individual Activity:

Assign an out-of-class activity.

    a. Students are to spend one week prior to the lesson looking in magazines, catalogues, and stores for articles relating to parabolic reflectors and their uses.

    b. Ask some of the students to briefly address the class concerning their findings about uses of parabolic reflectors.

    d. For extra credit, have the students bring in materials for the solar cookers. Also, if any students have made solar cookers, ask them to bring these for display purposes.

Cooperative Activity:

Use this activity as a class or team project:

    a. Have the students form teams to design and construct the solar cooker.

    b. Each team is to sketch and turn in a final design for a solar cooker.

    c. Construct a prototype if possible to determine the efficiency and focal point.

    d. Construct the final product. Test the product.

    e. Have a solar cookout and give out awards for best, fanciest, largest, smallest, and most innovative solar cookers. You might also have an award for the solar cooker made with the most recycled materials.

Overhead Transparency Masters can be found in Appendix A.

Chapter Review Questions and Answers can be found in Appendix B.

Crosstech Puzzles and Solutions can be found in Appendix C.

Chapter Evaluation Material can be found in Appendix D.

Technology Activity Guidelines can be found in Appendix E.

A General Safety Unit can be found in Appendix F.

POWER

Chapter 14, Text pages 400 to 423

Chapter Summary

Chapter 14, "Power", covers power, engines, transmissions, Newton's third law of motion, generators, and electric motors.

Chapter Outline

A. What is Power? (p. 401)
B. Power Systems (p. 402)
C. Engines (p. 402)
   1. External Combustion Engines
   2. Internal Combustion Engines
   3. Reaction Engines
   4. Nuclear Reactors
   5. Electric Motors
D. Transmissions (p. 410)
   1. Mechanical Power Transmission
   2. Hydraulic Power Transmission
   3. Electrical Power Transmission
E. Continuous versus Intermittent Power Systems (p. 413)
F. Summary (p. 414)
G. Review Questions (p. 417)
H. Crosstech
I. Activities (pp. 418–423)

Major Concepts:

■ Power is the amount of work done during a given period of time.

■ An engine is a machine that uses energy to create mechanical force and motion.

■ A transmission is a device that transmits force from one place to another or changes its direction.

■ Modern engines change the energy stored in fuel to mechanical force and motion.

■ Both external and internal combustion engines change the potential energy stored in a fuel into heat. The heat expands a gas, which moves a piston.

■ Newton's third law of motion states that for every action, there is an equal and opposite reaction.

■ A generator changes rotary motion into electrical energy.

■ An electric motor changes electrical energy into rotary motion.

Student Behavioral Objectives:

Upon completion of the "Troubleshooting" activity, the student should be able to:

1. Explain the three subsystems of a small engine: the ignition, compression, and fuel.

2. Tear down and rebuild the fuel, ignition, and compression subsystems of a small engine.

3. Perform simple tests to check a small engine to determine if ignition is occurring, if proper compression exists, and if the correct amount of fuel is being burned.

4. Correctly identify the parts of a simple engine.

5. Practice safe habits in working with engines. Use the correct tools for each operation.

6. Explain basic math and science concepts that relate to an engine.

Hints for Chapter 14

Instructional Suggestions:

a. Announce in advance the activity, materials needed, and outside-of-class assignments.
b. Ask students check at home, in junk yards, and at local engine repair shops for broken engines. Have them bring these to school.
c. Prepare and duplicate any information and instructions (especially that concerning safety in working with small engines) needed for the activity.
d. Invite a guest speaker and judges from engine repair shops to talk to the class and help judge the competitions.
e. Display cutaway models of two- and four-cycle engines.
f. Prepare lab equipment for the activity. Make sure safety instructions for engine repair are posted around the lab.

Power systems are discussed in Chapter 14, while transportation systems are discussed in Chapter 15. It may be helpful to combine these two discussions in some cases, although power systems are used in instances other than vehicles.

The introduction to this chapter continues the definitions and description of the relationship among work, energy, and power begun in Chapter 13. If a student's force expended in moving an object can be measured by a scale or other means for a period of time, then the power produced can be calculated by the class. It is always interesting to demonstrate what a small fraction of horsepower a typical student can produce. That amount of power can then be compared to the power produced by even small engines.

Under the definition used for an engine on page 402, windmills and other simple energy harnessers are engines. Frequently, these types of engines are used in a technology lab in problem solving or other activities involving the design of wind-powered vehicles, rubber-band-powered vehicles, or gravity-driven vehicles.

Although steam engines are generally treated as a matter of history, a discussion of their operation presents a good opportunity to demonstrate (or perhaps to construct a mechanism) how reciprocating motion is changed into rotary motion using a piston and flywheel. This technique is used in an internal combustion engine as well.

The idler pulley clutch mechanism shown on page 404, which was used during the Industrial Revolution to deliver power to individual machines from a common overhead rotating shaft, is also used on some go-carts, snow blowers, and other small engine machines. It may be possible to incorporate this kind of technology into a lab project.

A small model airplane engine may be used in a lab project as an example of an internal combustion engine. While this will demonstrate the principles, it should be noted that car engines have many other systems to carefully control fuel economy, optimize efficiency, automatically control speed, etc., and are therefore far more complex. Such an engine can be viewed as being a system made up of many smaller subsystems.

Reaction engines have replaced internal combustion engines in commercial aviation due to their lower maintenance requirements and operating costs, as well as the fact that they are more powerful. If desired, a discussion of jet engines can be coupled with the section on lift, drag, and thrust in Chapter 15.

The transmission component of a power system is treated on pages 410–413. For simplicity, transmissions are considered to be mechanical, hydraulic, pneumatic, electrical, or magnetic. Mechanical transmissions are probably the easiest to build, demonstrate, and visualize for students. Hydraulic transmissions make interesting activities, however, as operational model elevators using plastic tubing and other operational models of devices that students are familiar with but do not understand the principles behind.

A discussion of electric power transmission is a good opportunity to emphasize safety when working with electricity. Ordinary house voltage is enough to cause injury or death, and students should be warned of this. More dangerous, of course, are the voltages on utility poles; students should be warned of this danger and instructed never to climb utility poles or touch any string or wire that has come in contact with wires.

Hints for Activities

1. Have students bring in two- or four-cycle engines and repair these in the lab.
2. Get a salvage yard to donate broken two- and four-cycle engines to the lab. Have the students tear down these engines and repair them.
3. When all students are using the same type of engine, have troubleshooting competitions. Also have tear down competitions with 3 to 4 students per team. Have one member in charge of recording data, one in charge of tools, and two in charge of tear down.
4. Have teams bring in engines and repair and refinish the engine. Use individuals from local firms to judge the before photos and the finished product. A notebook with data, including photos of the engine before and after the tear down process, should be included. Have students describe the process followed for each subsystem. During the competition, one student should be the spokesperson to explain to the judges the extent of the problems, the process followed in repair, and costs of repair.

Individual Activity:

Assign an out-of-class activity.

a. Students are to spend one week prior to the lesson looking for broken engines at home, in their neighborhoods, in repair shops, and in junk yards. Emphasize that they need to get permission to get any of the engines.
b. Ask students to share with the class their experience with small engines in boats, lawn mowers, and go carts.
c. For extra credit, have the students bring in small engines they can get donated.
d. For extra credit, have a student take a broken engine and take it apart, labeling all parts of the three subsystems and making a display for the class.

Cooperative Activity:

Use this activity as a class or team project:

a. Assign the class to teams. Group the students so that all groups are equally filled according to abilities, etc.
b. Have the members of each team choose a team name.
c. Giving the students sets of instructions, have them run ignition, compression, and fuel tests. One person from each team is to record the data.
d. Set up two types of competitions. The first competition is a tear down and rebuild. All engines must be similar and all students must have access to the same types and numbers of tools. Using a stopwatch, time the process for the teams. The second type of competition is a troubleshooting competition. Run this one three times, each time having the team face a different type problem, which you, the teacher, have created. Tabulate the scores.
e. Award prizes for the winners and present certificates of completion to all teams.

Overhead Transparency Masters can be found in Appendix A.

Chapter Review Questions and Answers can be found in Appendix B.

Crosstech Puzzles and Solutions can be found in Appendix C.

Chapter Evaluation Material can be found in Appendix D.

Technology Activity Guidelines can be found in Appendix E.

A General Safety Unit can be found in Appendix F.

TRANSPORTATION

Chapter 15, Text pages 424 to 462

Chapter Summary

Chapter 15, "Transportation", describes the purpose and effects of transportation systems. The chapter also covers steam, modern, and intermodal transportation systems.

Chapter Outline

Major Concepts:

■ A transportation system is used to move people or goods from one location to another.

■ Modern transportation systems have helped to make countries interdependent.

■ The availability of rapid, efficient transportation systems has changed the way we live.

■ Transportation systems convert energy into motion.

- Steam was the first important source of mechanical power for transportation systems.
- Modern transportation systems often use internal combustion engines or electric motors.
- Intermodal transportation systems make optimum use of each type of transportation used in the system.
- Most transportation systems use vehicles to carry people or goods, but some systems do not use any vehicles.

Student Behavioral Objectives:

Upon completion of the "Air Flight" activity, the student should be able to:

1. Explain the difference between flights made within the earth's gravity and those made in microgravity conditions above the earth's atmosphere.
2. Design and use a variety of materials to construct an aircraft that will travel through the air.
3. Use various methods of launching the aircraft such as slings, rockets, throwing it by hand, and compressed air.
4. Be able to discuss fuel systems used in lower and upper atmospheric travel.

Hints for Chapter 15

Instructional Suggestions:

    a. Announce in advance the activity, materials needed, and outside-of-class assignments.
    b. Ask students to collect articles and brochures about air and space travel.
    c. Prepare and duplicate any information and instructions (especially that concerning safety in working with models, rockets, and compressed air) needed for the activity.
    d. Invite a guest speaker such as an astronaut or pilot to discuss flying and travel in space.
    e. Display models of aircraft and spacecraft.
    f. Prepare lab equipment for the activity. Make sure safety instructions regarding the use of rocket engines and tools and machines students will use are posted.

The early pages of this chapter discuss how history has been affected by the modes of transportation available at different times, and how modern transportation shapes the world we live in today. Modern, safe, rapid transportation of goods and people deeply affects our everyday lives, from our choices of places to live and work to our forms of entertainment.

Like other systems, transportation systems use the seven technological resources to produce outputs. One of the unique elements of transportation systems is the way in which they convert energy. The concepts of engine (motor) and transmission are universal in transportation systems. Most systems also have containers (vehicles) for the people or goods being transported, but there are some systems without vehicles.

Transportation systems are often very complex, containing many subsystems. Some of the subsystems of a large transportation system are shown in Chapter 3 (page 70). The interrelationship of communication systems, construction systems, manufacturing systems, management systems, and others can be seen in many large transportation systems.

The study of transportation systems has been divided into land, water, intermodal, air, space, and nonvehicle systems. While these categories can easily be studied separately, advances in technology have often affected several or all of them simultaneously. Examples include the development of the internal combustion engine (dooming steam power) and the application of the microprocessor to engine control and navigation.

There are many opportunities to explore problem solving and transportation systems together in the laboratory. Students can be challenged to build the fastest, safest (ability to carry a fragile cargo, such as an egg), strongest vehicle out of a limited set of resources. This kind of activity can be structured as a contest, with the students divided into groups. Different modes of transportation can be used for different classes, including water (often in a long thin tank), land (with mousetraps, rubber bands, CO_2 cartridges, sails driven by vacuum cleaners, etc. for "engines") and air (rockets, gliders, etc.).

In such projects, math and science principles can easily be brought into focus by having students experiment with shapes or different materials to find optimum solutions before building final products. For example: different materials can be tested for buoyancy before making boats; sails can be made in different shapes and tested before designing a boat and mast to hold them; wings can be tested in a wind tunnel made with a summer cooling fan, etc.

Nonvehicle transportation systems also make good subjects for laboratory projects. Conveyor belts or roller systems can be built that integrate with a manufacturing project of the same or another class, or students can be challenged to design and build a model of a moving sidewalk or escalator. An elevator using a hydraulic or pneumatic lift system (see Chapter 14) can also be built using cardboard and rubber hose.

Simple projects such as this illustrate the principles involved, allow students to build a project using them, and act as a vehicle for the instructor to point out recent developments in these seemingly "standard" technologies. (Example: in high rise buildings with many elevators that may be at many different floors at any given time, the elevators are controlled by microprocessors that calculate which elevator should respond to any request for elevator service, ensuring that the waiting time is kept to a minimum).

Hints for Activities

1. Have students bring in models of spacecraft and construct these.
2. Assign students to teams and direct them to design, construct, and present to the class for testing an aircraft designed to travel in space.
3. Have students develop notebooks on travel, using newspapers, magazines, brochures, and small posters.
4. Invite an astronaut or pilot to class to talk about flying. Or, invite a person with the rail system to talk about travel by train.
5. Hold competitions among the students for the fastest, highest flying, craziest stunt, best designed, and most colorful aircraft.

Individual Activity:

Assign an out-of-class activity.

    a. Students are to spend one week prior to the lesson researching and looking at models of airplanes and spacecraft.
    b. Ask students to share with the class their experiences with building model aircraft and spacecraft. Relate the types of, and uses for, the various models they have seen or constructed.
    c. For extra credit, have the students bring in models and suspend these from the ceiling.
    d. Have each student design and construct a vehicle that will fly through the air and out into space.

Cooperative Activity:

Use this activity as a class or team project:

    a. Divide the class into teams. Group the students so that all groups are equally filled according to abilities, etc.
    b. Have the members of each team choose a team name.
    c. Assign the teams to a brainstorming session where they design a space vehicle and plan how to construct it and which materials to use.
    d. Have the students construct the spacecraft and compile a notebook of their sketches, information, and flight data.
    e. Hold a competition. Ask a pilot or astronaut to judge the vehicles for flight, design, material usage, etc.
    f. Award prizes for the winners and present certificates of completion to all teams.

Overhead Transparency Masters can be found in Appendix A.

Chapter Review Questions and Answers can be found in Appendix B.

Crosstech Puzzles and Solutions can be found in Appendix C.

Chapter Evaluation Material can be found in Appendix D.

Technology Activity Guidelines can be found in Appendix E.

A General Safety Unit can be found in Appendix F.

IMPACTS FOR TODAY AND TOMORROW

Chapter 16, Text pages 464 to 493

Chapter Summary

Chapter 16, "Impacts For Today and Tomorrow", includes the outputs of a technological system, both positive and negative. The chapter covers environmental concerns and ergonomics in relationship to technology and society. Methods of predicting the future are also addressed.

Chapter Outline

A. Impacts of Technology (p. 465)
1. Should We Go Back to Nature?
2. Matching Technology to the Individual
3. Matching Technology to the Environment
B. Seeing Into the Future of Technology (p. 469)
C. Communication in the Future (p. 471)
1. Communication at Home
2. Communication at Work
3. Communication and the Consumer
D. Manufacturing in the Future (p. 474)
1. Manufacturing in Space
E. Construction in the Future (p. 476)
F. Transportation in the Future (p. 478)
1. Air Transportation
2. Space Transportation
3. Ground Transportation
G. Biotechnical Systems in the Future (p. 480)
1. Ethics
2. Food Production
H. Energy in the Future (p. 481)
I. Futuring—Forecasting New Technologies (p. 482)
1. Futuring Techniques
J. Summary (p. 485)
K. Conclusion (p. 486)
L. Review Questions (p. 487)
M. Crosstech (p. 487)
N. Activities (pp. 488–493)

Major Concepts:

■ Outputs of a technological system can be desired, undesired, expected, or unexpected.

■ People determine whether technology is good or bad by the way they use it.

■ Technology produces many positive outputs and solves many problems. Sometimes, however, negative outputs create new problems.

■ Technology must be fitted to human needs.

■ Technology must be adapted to the environment.

■ Existing technological systems will act together to produce new, more powerful technologies.

■ Using futuring techniques, people can anticipate the consequences of a new technology.

Student Behavioral Objectives:

Upon completion of the "Mapping the Future" activity, the student should be able to:

1. Complete and explain a futures wheel.

2. Describe the different ways of predicting the future and the advantages and disadvantages of each:
   a. Futures wheel
   b. Trend analysis
   c. Delphi survey
   d. Cross-impact analysis

3. As a member of a team, choose one method of predicting the future, use it, and report the results to the class.

Hints for Chapter 16

Instructional Suggestions:

   a. Announce in advance the activity, materials needed, and outside-of-class assignments.
   b. Assign outside activity of compiling information about the future in technology.
   c. Prepare and duplicate any information and instructions needed for the activity.
   d. Invite a guest speaker from industry to talk about the future in industry as they see it.
   e. Display in the room any models of future technological devices. Make a bulletin board of articles about the future in technology.

Ergonomics (human factors engineering) can provide a variety of laboratory activity ideas. Students may be asked to redesign a device so that it may be more easily used by a disabled person. Or, a computer table can be designed by students so that a human user is comfortably positioned. Arm, leg, and torso measurements of fellow students can be taken in order to ergonomically design a chair or desk. A good reference is *Ergonomics: Making Products and Places Fit People,* by Kathlyn Gray, Enslow Publishers, Inc., Box 777, Hillside, NJ 07205.

Students can be asked to redesign a technological device to make it better match human or environmental needs. A typewriter keyboard could be redesigned so that the strongest fingers are placed over the most frequently used keys; or students may be challenged to develop a system that would provide a warning signal if the indoor environment became too hot, too smoky, or too damp.

Future-oriented activities may involve the generation of futures wheels. Pick an event that is a likely future, such as the development of a cure for cancer, or the transport of private citizens into space. Use that event to forecast outcomes. Students may be asked to design and build future habitats in alien locations, like under the ocean or in outer space. Systems for human life support in these environments would have to be modeled.

Perhaps students would enjoy modeling transatmospheric aircraft, or a vehicle similar to the Planetran Express, which was designed to travel underground at a speed of 14,000 mph from coast to coast (see *Transplanetary Subway Systems — A Burgeoning Capability,* by Robert M. Salter (February, 1978), The Rand Corporation, 1700 Main Street, Santa Monica, CA 90406 (Rand/6092)

Because applications of biotechnology will no doubt increase considerably in the decades ahead, students may wish to research career opportunities in those fields. The Industrial Biotechnology Association (2115 East Jefferson Street, Rockville, MD 20852) publishes a host of introductory information on the subject, including a booklet entitled "Careers in Biotechnology." A 10-minute video entitled "The Chemical Engineer in Biotechnology"is available in 1/2" VHS format from the American Institute of Chemical Engineers, Career Guidance Division, 345 East 47 Street, New York City, NY 10017; (212) 705-7319.

Teachers may wish to contact the Cold Spring Harbor Laboratory, Bungtown Road, Cold Spring Harbor, NY 11724 (516-367-8397). Among its other endeavors, this famous research center has been developing hands-on activities in the area of biotechnology which are geared to high school students.

The University of Arizona's Environmental Research Labs, Tucson International Airport, Tucson, AZ 85706-6985 (602-626-2931), has been growing plants that can be irrigated with salt water (halophytes), as well as doing research in hydroponics. The director is Carl Hodges.

Hints for Activities

1. Divide the class into teams. Assign each team the same type of means to predict the future.
2. Compare and compile the results of the futuring technique used.
3. Divide the class into four teams. Assign each team one of the futuring methods. Each team is to use the method, construct a poster of the results, and submit to the class a report of their findings. The report should be done on a word processor.

Individual Activity:

Assign an out-of-class activity.

   a. Students are to spend one week prior to the lesson looking at television and reading newspapers and magazines for articles relating to the future in technology.

   b. Ask students to share with the class what they have seen and heard about the future in technology.

   c. For extra credit, have the students bring in futuristic models of airplanes, spacecrafts, robots, and other toys. Display these on a table.

   d. Have each student design and construct a model of a futuristic type device.

   e. Hold a competition for the most futuristic device.

Cooperative Activity:

Use this activity as a class or team project:

   a. Divide the class into teams. Group the students so that all groups are equally filled according to abilities, etc.

   b. Have the members of each team of experts choose a team name.

   c. Assign the teams to a brainstorming session; they must use the futuring techniques they were assigned.

   d. Each team is to make a poster and write a report (using a word processor) on the results of their use of the futuring technique.

   e. Hold a competition. Have individuals from industry judge the presentations of each team. Give out certificates.

Overhead Transparency Masters can be found in Appendix A.

Chapter Review Questions and Answers can be found in Appendix B.

Crosstech Puzzles and Solutions can be found in Appendix C.

Chapter Evaluation Material can be found in Appendix D.

Technology Activity Guidelines can be found in Appendix E.

A General Safety Unit can be found in Appendix F.

APPENDIX A TRANSPARENCY MASTERS

TECHNOLOGY IN A CHANGING WORLD

MAJOR CONCEPTS

■ Technology affects our routines.

■ Science is the study of why natural things happen the way they do.

■ Technology is the use of knowledge to turn resources into goods and services that society needs.

■ Science and technology affect all people.

■ People create technological devices and systems to satisfy basic needs and wants.

■ Technology is responsible for a great deal of the progress of the human race.

■ Technology can create both positive and negative social outcomes.

■ Combining simple technologies can create newer and more powerful technologies.

■ Technology has existed since the beginning of the human race, but it is growing at a faster rate than ever before.

TECHNOLOGY IN A CHANGING WORLD

KEY WORDS

Agricultural	Iron Age
Bronze Age	Manufacturing
Change	Mass production
Communications	Resources
Construction	Science
Energy	Stone Age
Exponential	Technology
Industrial Revolution	Technologically literate
Information Age	

TECHNOLOGY SATISFIES

Our Need to Produce Food

Our Medical Needs

Our Need for Manufactured Items

Our Need for Energy Sources

Our Need to Communicate

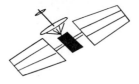

Our Transportation Needs

RESOURCES FOR TECHNOLOGY

MAJOR CONCEPTS

After reading this chapter, you will know that:

■ Every technological system makes use of seven types of resources.

■ The seven resources used in technological systems are people, information, materials, tools and machines, energy, capital and time.

■ Since there is a limited amount of certain resources on the earth, we must use resources wisely.

■ Solving technological problems requires skill in using all seven resources.

KEY WORDS

Capital	Machines
Coal	Material
Energy	Nuclear Energy
Finite	Oil
Gas	People
Geothermal	Resources
Hydroelectricity	Solar
Inclined plane	Synthetic
Information	Time
Laser	Tools

RESOURCES FOR TECHNOLOGY

1. **PEOPLE**

2. **INFORMATION**

3. **MATERIALS**

4. **TOOLS AND MACHINES**

5. **ENERGY**

6. **CAPITAL**

7. **TIME**

Renewable and Nonrenewable RAW MATERIALS

WOOD RUBBER   ANIMALS   PLANTS

Renewable
(Can be grown or replaced)

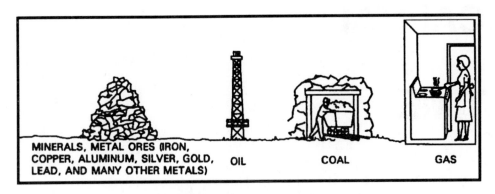

MINERALS, METAL ORES (IRON, COPPER, ALUMINUM, SILVER, GOLD, LEAD, AND MANY OTHER METALS)   OIL   COAL   GAS

Nonrenewable
(Cannot be grown or replaced)

SIX SIMPLE MACHINES

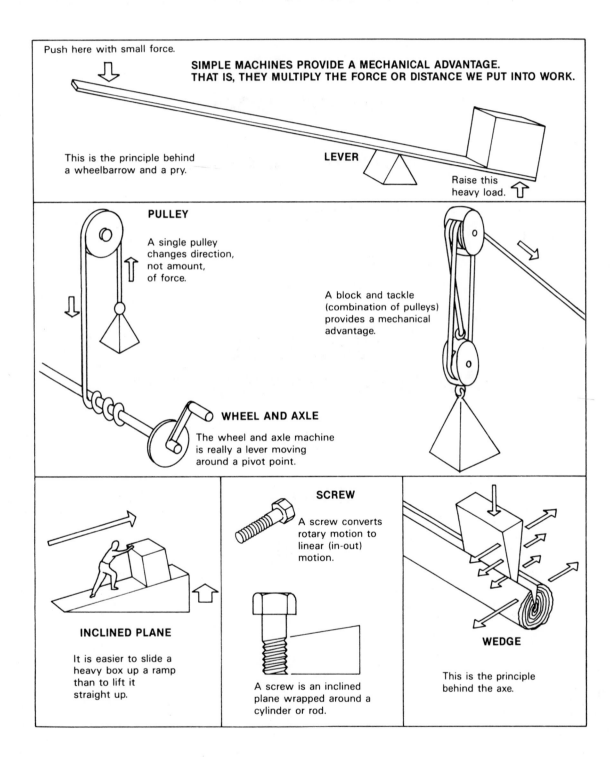

Push here with small force.

SIMPLE MACHINES PROVIDE A MECHANICAL ADVANTAGE. THAT IS, THEY MULTIPLY THE FORCE OR DISTANCE WE PUT INTO WORK.

This is the principle behind a wheelbarrow and a pry.

LEVER

Raise this heavy load.

PULLEY

A single pulley changes direction, not amount, of force.

A block and tackle (combination of pulleys) provides a mechanical advantage.

WHEEL AND AXLE

The wheel and axle machine is really a lever moving around a pivot point.

INCLINED PLANE

It is easier to slide a heavy box up a ramp than to lift it straight up.

SCREW

A screw converts rotary motion to linear (in-out) motion.

A screw is an inclined plane wrapped around a cylinder or rod.

WEDGE

This is the principle behind the axe.

RESOURCES CAN BE EITHER:

LIMITED
In short supply

Water in some places

Petroleum

UNLIMITED
Available in great quantity

Clay

Iron ore

Sand

TIME MEASUREMENT THROUGHOUT THE AGES

Agricultural Age

Sundials or burning rope were used.

Middle Ages

Water clocks were used.

Industrial Age

Time was told using mechanical clocks.

Information Age

Today, quartz and atomic clocks were used.

PROBLEM SOLVING AND SYSTEMS

MAJOR CONCEPTS

After reading this chapter, you will know that:

◼ People must solve problems that involve the environment, society, and the individual.

◼ A carefully thought out multi-step procedure is the best way to solve problems.

◼ A good solution often requires making trade-offs.

◼ Technological decisions must take both human needs and the protection of the environment into consideration.

◼ People design technological systems to satisfy human needs and wants.

◼ All systems have inputs, a process, and outputs.

◼ The basic system model can be used to analyze all kinds of systems.

◼ Feedback is used to make the actual result of a system come as close as possible to the desired result.

◼ Systems often have several outputs, some of which may be undesirable.

◼ Subsystems can be combined to produce more powerful systems.

PROBLEM SOLVING AND SYSTEMS

KEY WORDS

Alternatives	Model
Actual results	Monitoring
Basic research	Open-loop system
Basic system model	Optimization
Brainstorming	Output
Control	Problem solving
Closed-loop system	Process
Design folder	Prototype
Design brief	Research
Desired results	Subsystem
Feedback	System
Input	Trade-off
Insight	Trial and error
Market research	

THE TECHNOLOGICAL METHOD OF SOLVING A PROBLEM

Step 1

DESCRIBE THE PROBLEM

Step 2

DESCRIBE THE RESULTS YOU WANT

Step 3

GATHER INFORMATION

Step 4

THINK OF ALTERNATIVE SOLUTIONS

Step 5

CHOOSE THE BEST SOLUTION

Step 6

IMPLEMENT THE SOLUTION

Step 7

EVALUATE THE SOLUTION AND MAKE CHANGES

A BASIC TECHNOLOGICAL SYSTEM

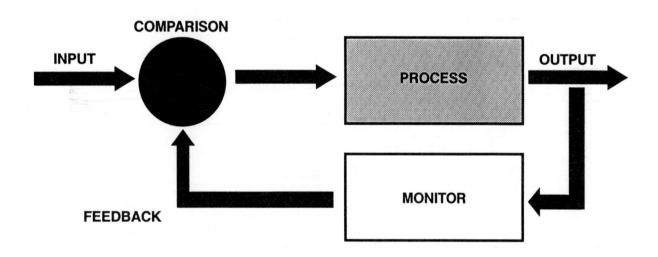

INPUT: Is the command we give a system (desired result).

PROCESS: Is the action part of a technological system.

OUTPUT: Is the actual result obtained from a system.

FEEDBACK: Is the use of information about the output of a system.

SYSTEM DIAGRAM FOR PROBLEM SOLVING

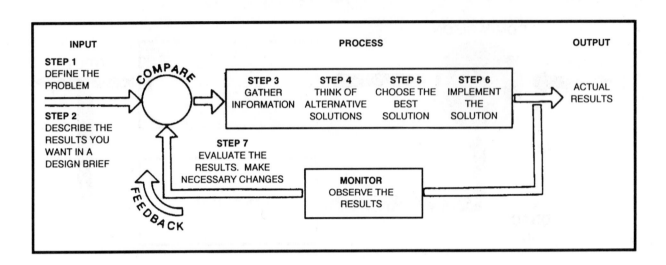

The problem-solving system has the same system diagram as technological systems.

SEVEN TECHNOLOGICAL RESOURCES COMBINED TO PROVIDE AN OUTPUT

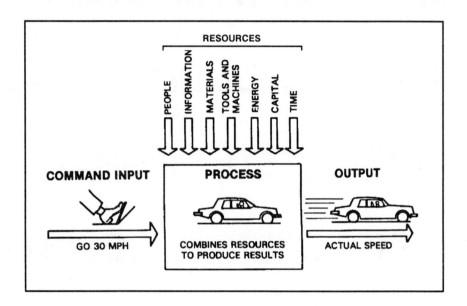

The process combines the seven technological resources to produce the desired result.

The output of this system is the speed that the car actually goes.

DRIVING A CAR

FOUR KINDS OF OUTPUT

1. EXPECTED, DESIRABLE

2. EXPECTED, UNDESIRABLE

3. UNEXPECTED, DESIRABLE

4. UNEXPECTED, UNDESIRABLE

SYSTEM MODEL OF THE HUMAN BODY

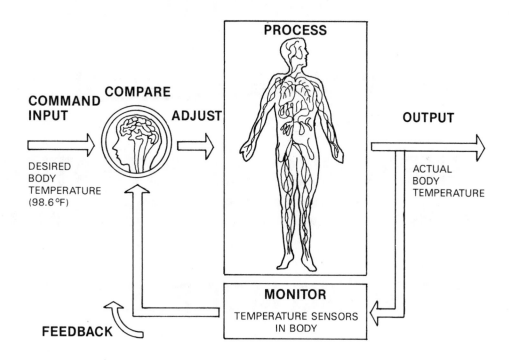

System model of the human body. Temperature sensors in the body provide the feedback needed to maintain the desired body temperature.

MODELING TECHNIQUES

1. CHARTS AND GRAPHS

2. MATHEMATICAL MODELS

3. SKETCHES, ILLUSTRATIONS, AND TECHNICAL DRAWINGS

4. WORKING MODELS

5. COMPUTER SIMULATION

| CHAPTER 4 | # THE ELECTRONIC COMPUTER AGE |

MAJOR CONCEPTS

After reading this chapter, you will know that:

■ The use of electronics has revolutionized all aspects of technology.

■ Electric current is the flow of electrons through a conductor.

■ Electronic circuits are made up of components. Each component has a specific function in the circuit.

■ Integrated circuits are complete electronic circuits made at one time on a piece of semiconductor material.

■ Computers are general-purpose tools of technology.

■ Computers use 1s and 0s to represent information.

■ Computers have inputs, processors, outputs, and memories.

■ Computers operate under a set of instructions, called a program. The program can be changed to make the computer do another job.

■ Computers can be used as systems or can be small parts of larger systems.

71

THE ELECTRONIC COMPUTER AGE

KEY WORDS

Ampere	Integrated circuit
Analog	Memory
Bit	Operating system
Byte	Printed circuit board
Circuit	Printer
Component	Processor
Compounds	Program
Conductor	Random access memory (RAM)
Current	Resistance
Digital	Semiconductor
Electron	Supercomputer
I/O	Transistor
Insulator	Voltage

ELECTRIC CURRENT FLOW

The flow of electrical current through a wire is simialr in concept to the flow of water through a thin pipe. In an electric circuit, the electromotive force might be supplied by a battery (A). In the water pipe, the force might be applied by a person pushing a piston (B).

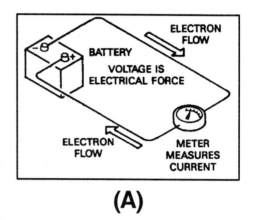

(A)

In both cases, a greater force will make a larger current flow. With the same force, if the water pipe is made bigger in diameter, it will offer less resistance to the water, and more water will flow (C). In a similar way, if a larger wire is used, it will offer less resistance to electron flow, and more electrical current will flow.

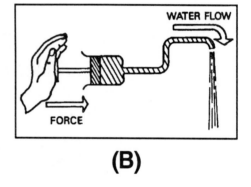

(B)

The equation that describes current flow in an electrical circuit is called Ohm's Law. It states that

$$\text{Current (amps)} = \frac{\text{Voltage (volts)}}{\text{Resistance (ohms)}}$$

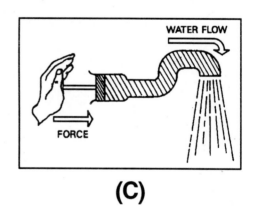

(C)

If voltage (force) gets larger, current gets larger. If resistance gets larger, current gets smaller.

COMPUTER SYSTEMS

PROCESSOR
Heart of the computer

MEMORY
Place the program is stored

INPUT
Information given to computer

OUTPUT
Screen display or hard copy

COMPUTER
INPUT DEVICES

KEYBOARD AND TAPES

LIGHT SENSITIVE SCREENS

HARD AND FLOPPY DISKS

OPTICAL CHARACTER READER

SPEECH RECOGNITION

MOUSE AND LIGHT PENS

COMPUTER
OUTPUT DEVICES

VIDEO MONITOR
(Cathode Ray Tube Screen)

PRINTERS — HARD COPY
(Daisy wheel and dot matrix)

AUDIO OUTPUT
(Tones, beeps, music, voice)

TYPES OF COMPUTERS

MICROCOMPUTERS
cars, appliances, and computers

MINICOMPUTERS
used by small companies

MAINFRAME
Used by universities, large companies, and government agencies

SUPERCOMPUTERS
Used for research

SOFTWARE

OPERATING SYSTEM
Allows user to control and access memory, printer, and peripheral devices.

MSDOS

OS/2

APPLICATIONS PROGRAM
Tells computer to carry out specific tasks.

Games

Word processors

Car engine control programs

PROGRAMMING LANGUAGES
Pascal

BASIC

SYSTEMS DIAGRAM APPLIED TO COMPUTERS

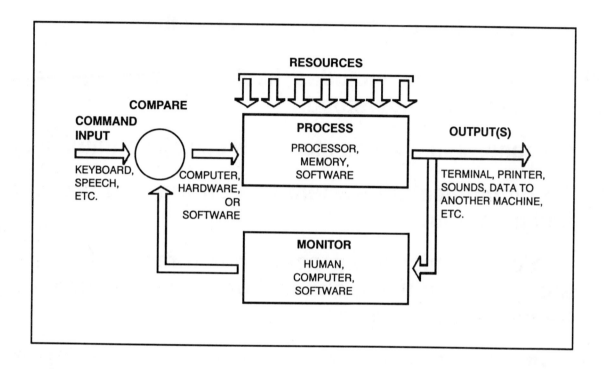

CHAPTER 5

COMMUNICATION SYSTEMS

MAJOR CONCEPTS

After reading this chapter, you will know that:

■ Communication includes having a message sent, received, and understood.

■ Humans, animals, and machines can all communicate.

■ A communication system has an input, a process, and an output.

■ The communication process has three parts. They are: a transmitter, a channel on which the message travels, and a receiver.

■ Communication systems are used to inform, persuade, educate, and entertain.

■ Communication systems require the use of the seven technological resources.

■ Two kinds of communication systems are graphic and electronic.

COMMUNICATION SYSTEMS

KEY WORDS

Channel

Communication system

Graphic communication

Educate

Electronic communication

Entertain

Inform

Noise

Machine communication

Mass media

Persuade

Receiver

Transmitter

TYPES OF COMMUNICATIONS

PERSON-TO-PERSON
(Speech and expressions)

ANIMAL
(Humans with pets)

MACHINE
(Machine to machine or machine to human)

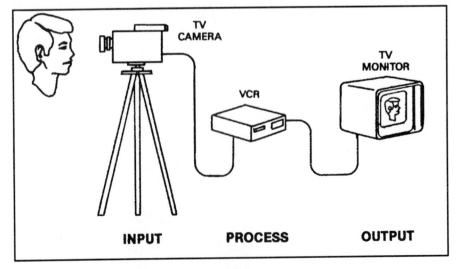

GENERAL SYSTEM DIAGRAM FOR A COMMUNICATION SYSTEM

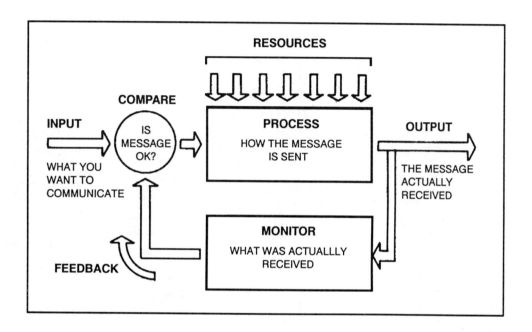

A RADIO COMMUNICATION SYSTEM

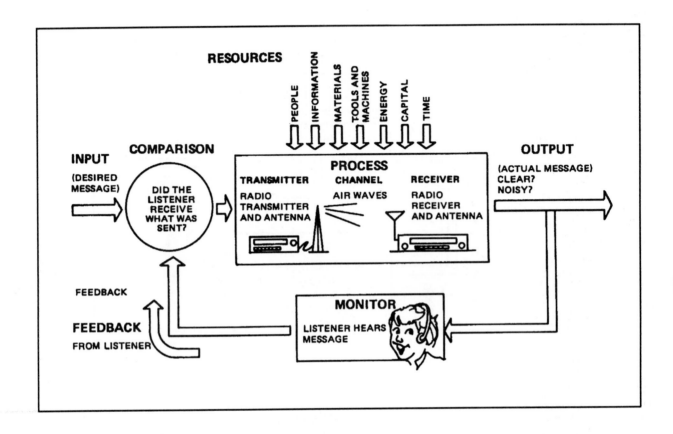

RESOURCES

PEOPLE
INFORMATION
MATERIALS
TOOLS AND MACHINES
ENERGY
CAPITAL
TIME

INPUT
(DESIRED MESSAGE)

COMPARISON

DID THE LISTENER RECEIVE WHAT WAS SENT?

PROCESS

TRANSMITTER
RADIO TRANSMITTER AND ANTENNA

CHANNEL
AIR WAVES

RECEIVER
RADIO RECEIVER AND ANTENNA

OUTPUT
(ACTUAL MESSAGE)
CLEAR?
NOISY?

FEEDBACK

FEEDBACK
FROM LISTENER

MONITOR
LISTENER HEARS MESSAGE

84

THREE PARTS OF A COMMUNICATION PROCESS

TRANSMITTING
(Sending the message)

Computer with modem

CHANNEL
(Route message takes)

Telephone wires

RECEIVER
(Accepts the message)

Computer, Monitor, and Modem

COMMUNICATION PROCESS:

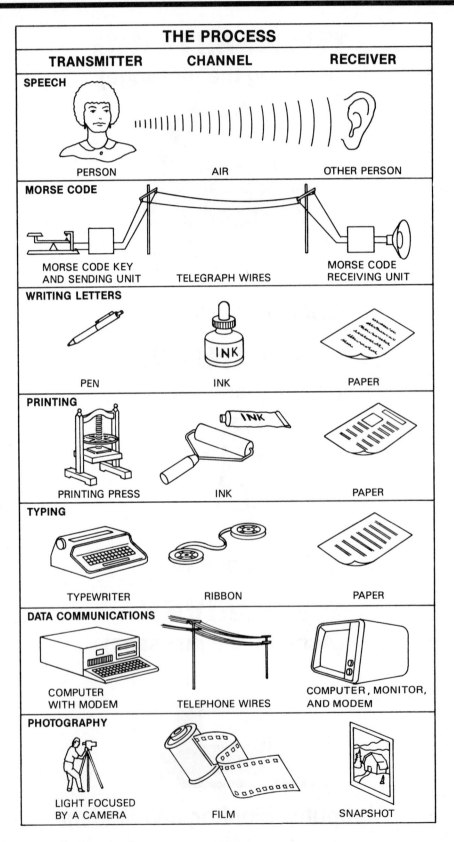

THE PROCESS

TRANSMITTER	CHANNEL	RECEIVER
SPEECH		
PERSON	AIR	OTHER PERSON
MORSE CODE		
MORSE CODE KEY AND SENDING UNIT	TELEGRAPH WIRES	MORSE CODE RECEIVING UNIT
WRITING LETTERS		
PEN	INK	PAPER
PRINTING		
PRINTING PRESS	INK	PAPER
TYPING		
TYPEWRITER	RIBBON	PAPER
DATA COMMUNICATIONS		
COMPUTER WITH MODEM	TELEPHONE WIRES	COMPUTER, MONITOR, AND MODEM
PHOTOGRAPHY		
LIGHT FOCUSED BY A CAMERA	FILM	SNAPSHOT

CATEGORIES OF A COMMUNICATION SYSTEM

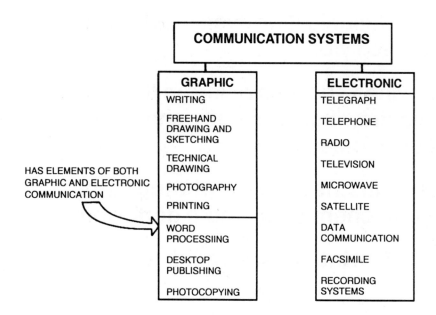

COMMUNICATION SYSTEMS

GRAPHIC	ELECTRONIC
WRITING	TELEGRAPH
FREEHAND DRAWING AND SKETCHING	TELEPHONE
TECHNICAL DRAWING	RADIO
PHOTOGRAPHY	TELEVISION
PRINTING	MICROWAVE
WORD PROCESSIING	SATELLITE
DESKTOP PUBLISHING	DATA COMMUNICATION
PHOTOCOPYING	FACSIMILE
	RECORDING SYSTEMS

HAS ELEMENTS OF BOTH GRAPHIC AND ELECTRONIC COMMUNICATION

87

GRAPHIC COMMUNICATION

MAJOR CONCEPTS

After reading this chapter, you will know that:

■ In a graphic communication system, the channel carries images or printed words.

■ Pictorial drawings show an object in three dimensions. Orthographic drawings generally show top, front, and side views of an object.

■ The five elements needed for photography are light, film, a camera, chemicals, and a darkroom.

■ Four types of printing are relief, gravure, screen, and offset.

■ Word processing improves office productivity.

■ Desktop publishing systems combine words and pictures.

CHAPTER 6

GRAPHIC COMMUNICATION

KEY WORDS

CAD

Daisy wheel

Desktop publishing

Dot matrix

Gravure printing

Isometric

Laser printer

Negative

Offset printing

Orthographic

Perspective

Phototypesetter

Pictorial

Relief printing

Screen printing

Technical drawing

GRAPHIC COMMUNICATION SYSTEMS

1. **WRITING**

2. **FREEHAND SKETCHING AND DRAWING**

3. **TECHNICAL DRAWING**

4. **PHOTOGRAPHY**

5. **PRINTING**

6. **PHOTOCOPYING**

7. **COMPUTER PRINTING**
   A. Word processing
   B. Desktop publishing

PICTORIAL DRAWINGS

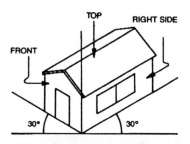

ISOMETRIC

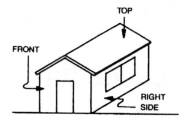

OBLIQUE

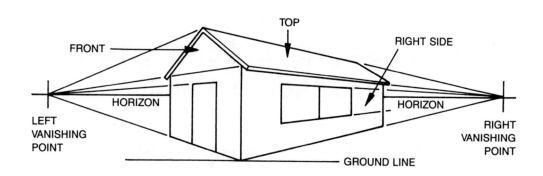

PERSPECTIVE

ADVANTAGES OF A COMPUTER-AIDED DESIGN SYSTEM

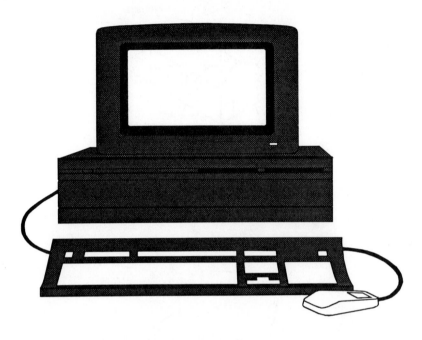

Saves time

Reduces chance of error

Drawings can be done faster

New designs and changes are quicker to make

Accurate and consistent

FIVE ELEMENTS OF PHOTOGRAPHY

1. **LIGHT**

2. **FILM**

3. **CAMERA**

4. **CHEMICALS**

5. **DARK AREA**

FOUR TYPES OF CAMERAS

VIEW CAMERA

VIEWFINDER CAMERA

TWIN LENS REFLEX

SINGLE LENS REFLEX

ELECTRONIC COMMUNICATION

MAJOR CONCEPTS

After reading this chapter, you will know that:

■ In an electronic communication system, the channel carries an electronic or electromagnetic signal.

■ Electricity and electronics have greatly changed communication technology.

■ The telephone is the most commonly used form of electronic communication.

■ In radio communication, the message is sent through the air from a transmitting antenna to a receiving antenna some distance away.

■ Computing and communication technologies are being used together to make powerful new tools for working with information.

■ A number of computers or computer devices joined together is called a network.

| CHAPTER 7 | ELECTRONIC COMMUNICATION |

KEY WORDS

Amplifier

ASCII

Broadcast

Cassette

Compact disk

Data communication

Distributed computing

Downlink

Electromagnetic wave

Fiber optic

Frequency

Local area network

Modem

Network

Office automation

Stereo

Tape head

Telecommute

Teleconference

Telegraph

Telephone

Tracks

Uplink

Wavelength

TRANSMITTING AND RECEIVING SYSTEMS

Telegraph

Telephone

Radio

Television

Microwave Communication

Satellite Communication

Fiber Optic Communications

DATA COMMUNICATION SYSTEMS

CENTRALIZED COMPUTING
All processing is done in one computer
with terminals and printers attached.

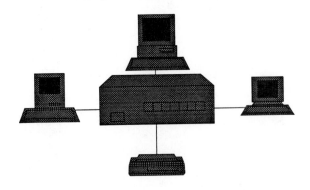

DISTRIBUTED COMPUTING
Small computers from other locations can send
larger, more difficult jobs to a large computer.

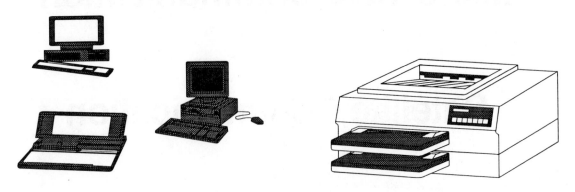

MODEMS

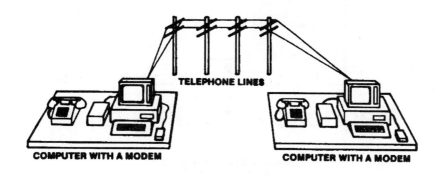

TELEPHONE LINES

COMPUTER WITH A MODEM   COMPUTER WITH A MODEM

MODEM. . .is a device used to send data signals over analog communication channels such as telephone circuits. (MODulator-DEModulator)

MODULATION. . .turning data signals into audio signals.

DEMODULATION. . .turning audio signals into data signals.

RECORDING SYSTEMS

Magnetic recording tape

Magnetic recording disk

Phonograph records

Optical disk

MODERN TYPES OF COMMUNICATION SERVICES

1. **TELECONFERENCING**

2. **TELECOMMUTING**

3. **ELECTRONIC BANKING**

4. **TWO-WAY CABLE TV**

<table>
<tr><td>CHAPTER
8</td><td></td></tr>
</table>

PROCESSING MATERIALS

MAJOR CONCEPTS

After reading this chapter, you will know that:

■ Processing materials involves changing their form to make them more useful to people.

■ The changing, or conversion, of resources occurs within the technological process.

■ Technological systems convert raw materials into basic industrial materials. Basic industrial materials are then converted into end products.

■ There are several ways of processing materials. These processes are forming, separating, combining, and conditioning.

■ Computers can control machines used to process materials.

■ Materials are then chosen on the basis of their mechanical, electrical, magnetic, thermal, or optical properties.

PROCESSING MATERIALS

KEY WORDS

Brittle

Casting

Ceramics

Coating

Combining

Composites

Compression

Conditioning

Conductor

Drilling

Ductile

Elastic

Extruding

Fastening

Ferrous metals

Forging

Forming

Gluing

Grinding

Heat-treating

Industrial materials

Insulator

Plasticity

Polymers

Pressing

Properties of materials

Raw materials

Recycle

Sawing

Separating

Shaping

Shearing

Tension

Thermal

Thermoplastics

Thermoset plastics

Torsion

Toughness

Turning

INDUSTRIAL MATERIALS

WOOD
(Hardwood, softwood, & manufactured board)

METAL
(Ferrous & nonferrous)

CERAMICS
(from clay or inorganic materials, plaster, cement, limestone, & glass)

PLASTIC
(Thermoplastic & Thermoset plastics)

FOUR WAYS OF PROCESSING MATERIALS

1. **FORMING. . . .casting, blow molding, vacuum forming**

2. **SEPARATING. . . .shearing, sawing, drilling, shaping**

3. **COMBINING. . . fastening, gluing, composites**

4. **CONDITIONING. . . .firing, heat treating, magnetizing**

FORMING PROCESSES

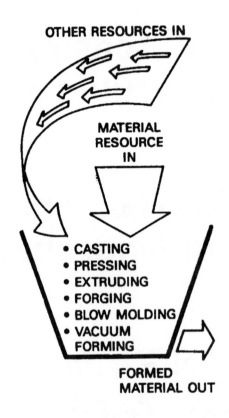

106

FORMING PROCESSES

1. CASTING. . .(made from molds)

2. pressing. . .material is pressed into mold

3. FORGING. . .metal is heated and pressed

4. EXTRUDING. . .softened material is heated and squeezed through opening

FORMING PROCESSES

5. **BLOW MOLDING. . .air blows softened plastic into molds (plastic bottle)**

6. **VACUUM FORMING. . .a vacuum draws heated and softened plastic over a product (blister packaging)**

SEPARATING PROCESSES

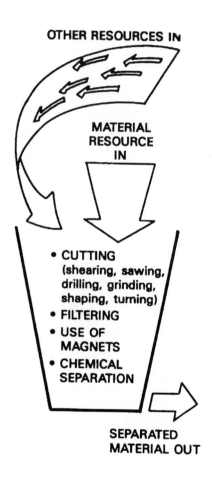

OTHER RESOURCES IN

MATERIAL
RESOURCE
IN

- CUTTING
  (shearing, sawing,
  drilling, grinding,
  shaping, turning)
- FILTERING
- USE OF
  MAGNETS
- CHEMICAL
  SEPARATION

SEPARATED
MATERIAL OUT

SEPARATING PROCESSES

CUTTING MACHINES

TURNING

DRILLING

**SHAPING &
PLANING**

**GRINDING &
SANDING**

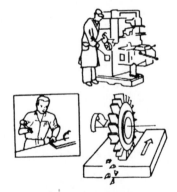

**MILLING &
SAWING**

SHEARING

SAWS AND SAWING

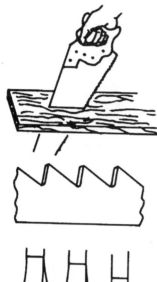

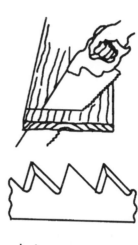

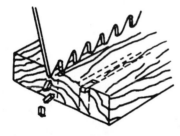

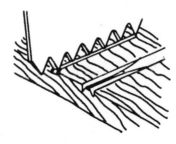

A RIPSAW IS USED TO CUT WITH THE GRAIN.
RIPSAW

A CROSSCUT SAW IS USED TO CUT
ACROSS THE GRAIN.
CROSSCUT SAW

COPING SAW

DOVETAIL SAW

KEYHOLE SAW

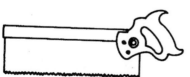

BACKSAW

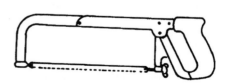

HACKSAW

111

COMBINING PROCESSES

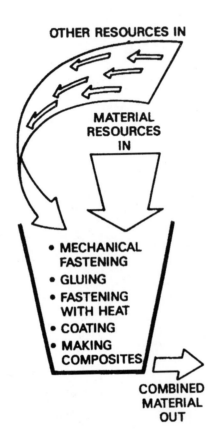

OTHER RESOURCES IN

MATERIAL
RESOURCES
IN

- MECHANICAL
  FASTENING
- GLUING
- FASTENING
  WITH HEAT
- COATING
- MAKING
  COMPOSITES

COMBINED
MATERIAL
OUT

COMBINING PROCESSES

GLUING

COATING

MECHANICAL FASTENING

FASTENING WITH HEAT

MAKING COMPOSITES

JIGS & FIXTURES

JIG. . .used to hold and guide the item being processed.

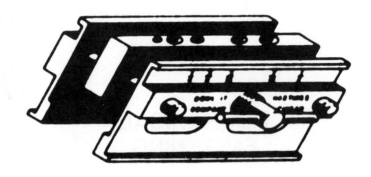

FIXTURE. . .used to keep the item being processed in the proper position.

TYPES OF NAILS

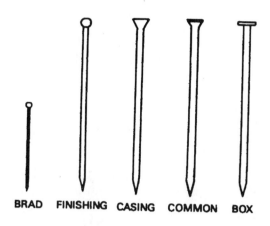

BRAD FINISHING CASING COMMON BOX

RELATIVE SIZES OF NAILS

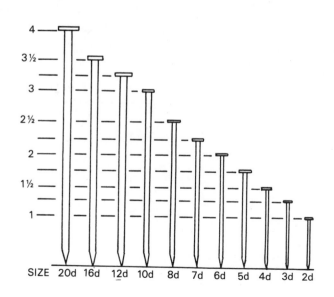

SIZE 20d 16d 12d 10d 8d 7d 6d 5d 4d 3d 2d

115

TYPES OF NAILS

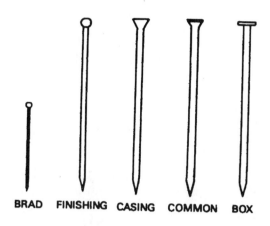

BRAD FINISHING CASING COMMON BOX

RELATIVE SIZES OF NAILS

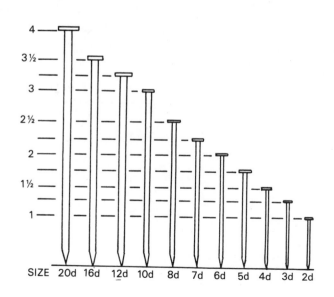

SIZE 20d 16d 12d 10d 8d 7d 6d 5d 4d 3d 2d

115

Copyright © 1993 by Delmar Publishers, Inc.®

TYPES OF SCREWS

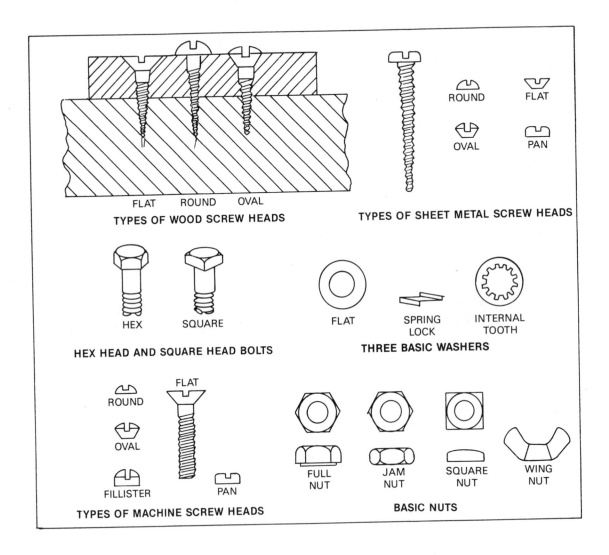

FLAT   ROUND   OVAL

TYPES OF WOOD SCREW HEADS

ROUND   FLAT

OVAL   PAN

TYPES OF SHEET METAL SCREW HEADS

HEX   SQUARE

HEX HEAD AND SQUARE HEAD BOLTS

FLAT   SPRING LOCK   INTERNAL TOOTH

THREE BASIC WASHERS

ROUND

OVAL

FLAT

FILLISTER   PAN

TYPES OF MACHINE SCREW HEADS

FULL NUT   JAM NUT   SQUARE NUT   WING NUT

BASIC NUTS

CONDITIONING PROCESSES

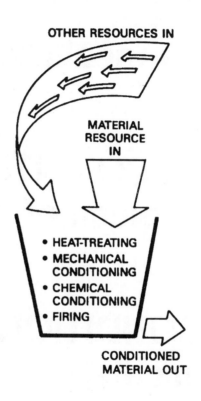

OTHER RESOURCES IN

MATERIAL
RESOURCE
IN

- HEAT-TREATING
- MECHANICAL
  CONDITIONING
- CHEMICAL
  CONDITIONING
- FIRING

CONDITIONED
MATERIAL OUT

CONDITIONING PROCESSES

HEAT TREATING. . .(hardening, tempering, & annealing)

MECHANICAL CONDITIONING. . . (hammering a piece of metal)

CHEMICAL CONDITIONING. . . (mixing chemicals)

MAGNETIZE. . .(making a magnet from a piece of steel by lining up molecules)

PROPERTIES OF MATERIALS

MECHANICAL

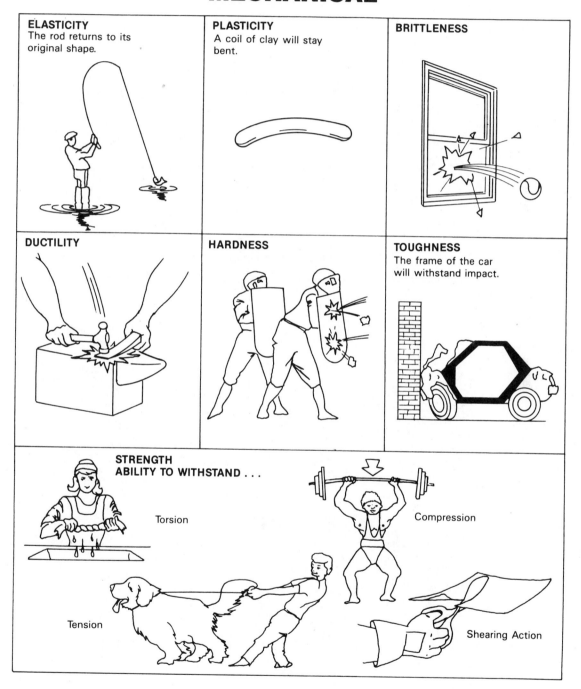

ELASTICITY
The rod returns to its original shape.

PLASTICITY
A coil of clay will stay bent.

BRITTLENESS

DUCTILITY

HARDNESS

TOUGHNESS
The frame of the car will withstand impact.

STRENGTH
ABILITY TO WITHSTAND . . .

Torsion

Compression

Tension

Shearing Action

119

PROPERTIES OF MATERIALS

OPTICAL

OPTICAL CLARITY

REFLECTIVITY

THERMAL

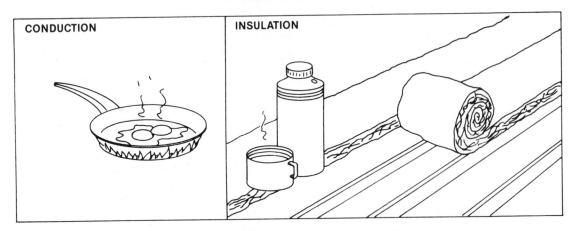

CONDUCTION

INSULATION

ELECTRICAL

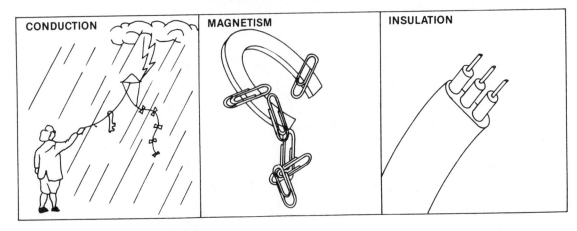

CONDUCTION

MAGNETISM

INSULATION

CHAPTER 9
MANUFACTURING

MAJOR CONCEPTS

After reading this chapter, you will know that:

■ Production technologies fill many people's needs and wants by means of manufacturing and construction systems.

■ Manufacturing is making goods in a workshop or factory. Construction is building a structure on a site.

■ Mass production and the factory system brought prices down. People were able to improve their standard of living.

■ Manufacturing systems make use of the seven types of technological resources.

■ There are two subsystems within the manufacturing system: the material processing system and the business and management system.

■ Automation has greatly improved productivity in manufacturing.

■ Computers and robots have improved product quality while bringing manufacturing costs down.

MANUFACTURING

KEY WORDS

Assembly line	Just-In-Time manufacturing
Automation	Manufacturing
CAD/CAM	Mass production
CIM	Numerical control
Custom-made	Production
Entrepreneur	Prototype
Factory	Quality control
Feedback control	Robot
Flexible manufacturing	Uniformity
Interchangeable	Union
Inventor	

ENTREPRENEUR

A person who comes up with a good idea and uses it to make money.

INVENTOR

An inventor comes up with totally new ideas.

INNOVATOR

An entrepreneur who improves an invention and leads to other uses.

MANUFACTURING SYSTEM DIAGRAM

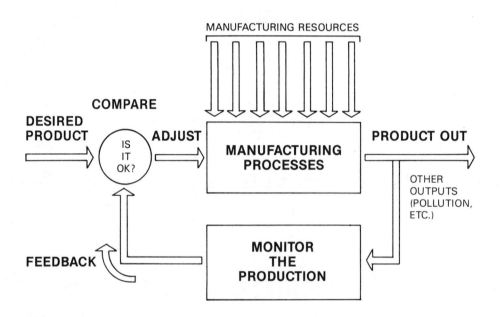

MANUFACTURING RESOURCES

COMPARE

DESIRED
PRODUCT

IS
IT
OK?

ADJUST

MANUFACTURING
PROCESSES

PRODUCT OUT

OTHER
OUTPUTS
(POLLUTION,
ETC.)

FEEDBACK

MONITOR
THE
PRODUCTION

RESOURCES FOR MANUFACTURING SYSTEMS

TOOLS & MACHINES

PEOPLE

MATERIAL

ENERGY

INFORMATION

CAPITAL

$

TIME

125

FLOWCHART
MAKING OF A METAL
SCREWDRIVER

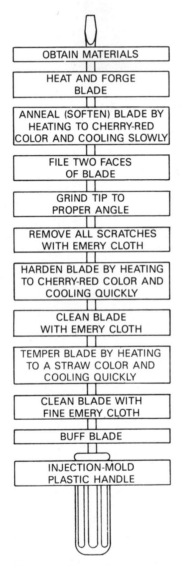

OBTAIN MATERIALS

HEAT AND FORGE
BLADE

ANNEAL (SOFTEN) BLADE BY
HEATING TO CHERRY-RED
COLOR AND COOLING SLOWLY

FILE TWO FACES
OF BLADE

GRIND TIP TO
PROPER ANGLE

REMOVE ALL SCRATCHES
WITH EMERY CLOTH

HARDEN BLADE BY HEATING
TO CHERRY-RED COLOR AND
COOLING QUICKLY

CLEAN BLADE
WITH EMERY CLOTH

TEMPER BLADE BY HEATING
TO A STRAW COLOR AND
COOLING QUICKLY

CLEAN BLADE WITH
FINE EMERY CLOTH

BUFF BLADE

INJECTION-MOLD
PLASTIC HANDLE

This flowchart shows the steps in making a metal screwdriver with a plastic handle.

SYSTEM DIAGRAM
AUTOMATICALLY
CONTROLLED
DRILL PRESS

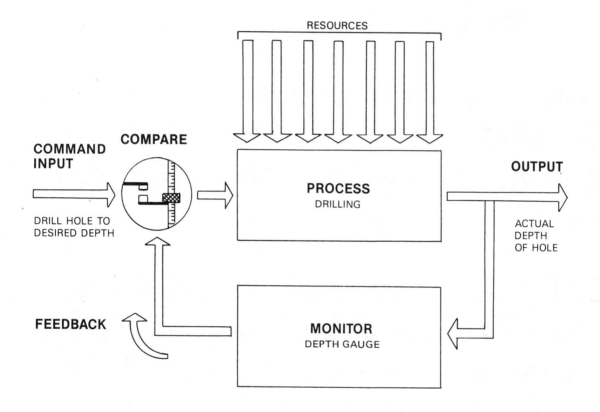

RESOURCES

COMMAND
INPUT

COMPARE

DRILL HOLE TO
DESIRED DEPTH

PROCESS
DRILLING

OUTPUT

ACTUAL
DEPTH
OF HOLE

FEEDBACK

MONITOR
DEPTH GAUGE

COMPUTERS IN MANUFACTURING

CAD. . .(Computer-Aided-Design) Drawings and designs are done and stored on computer.

CAM. . .(Computer-Aided Manufacturing) Computers are used to control machines on the factory floor.

CAD/CAM. . .Part is designed, tested, and stored as data. Numerical data is read by a computer-controlled machine which produces the part.

CIM. . .(Computer-Integrated Manufacturing) Computers store drawings and designs, data about parts, materials, inventories, bills, accounting, and scheduling of purchases.

COMPUTER INTEGRATED MANUFACTURING

CIM combines manufacturing, design, and business functions of a company under the control of a computer system.

FLEXIBLE MANUFACTURING. . .
Small batches of custom-tailored products are made by using production lines with reprogrammable machines.

JUST-IN-TIME MANUFACTURING. . .
Raw materials and purchased parts arrive at the factory just in time to be used on the production line. Product is shipped as soon as completed.

CHAPTER 10

CONSTRUCTION

MAJOR CONCEPTS

After reading this chapter, you will know that:

■ Construction refers to producing a structure on a site.

■ A construction system combines resources to provide a structure as an output.

■ Three subsystems within the construction system are designing, managing, and building.

■ Construction sites must be chosen to fit in with the needs of people and the environment.

■ A foundation is built to support a structure.

■ The usable part of a structure is called the superstructure.

■ Structures include bridges, buildings, dams, harbors, roads, towers, and tunnels.

CHAPTER 10

CONSTRUCTION

KEY WORDS

Arch	General contractor
Architect	Macadam
Cement	Mortgage
Concrete	Prefabricate
Construction	Specifications
Engineer	Steel
Estimator	Structure
Forms	Superstructure
Foundation	Surveyor

RESOURCES FOR CONSTRUCTION SYSTEMS

PEOPLE

Architect   Land Owner   Engineer
General Contractor   Craftspeople

INFORMATION

Plans   Specifications

MATERIALS

Concrete   Steel   Wood

TOOLS AND MACHINES

Hand Tools   Machine Tools
Cranes   Bulldozers

ENERGY

For operating equipment and support
the construction industry

CAPITAL

Money used for loans, rentals, and materials

TIME

Measurement of days to complete construction

SYSTEM DIAGRAM FOR CONSTRUCTION

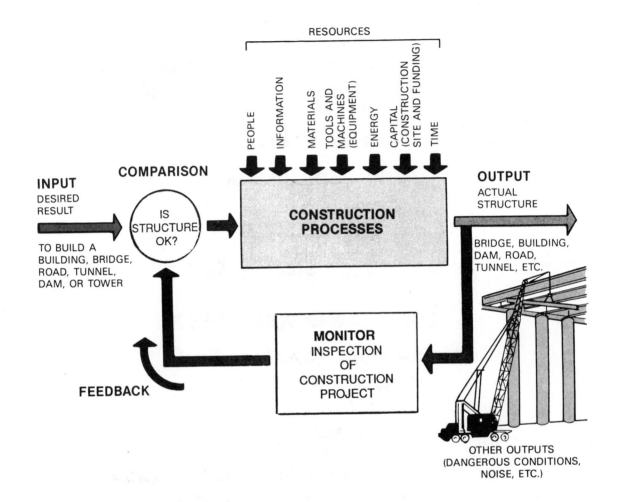

RESOURCES

PEOPLE

INFORMATION

MATERIALS

TOOLS AND
MACHINES
(EQUIPMENT)

ENERGY

CAPITAL
(CONSTRUCTION
SITE AND FUNDING)

TIME

INPUT
DESIRED
RESULT

TO BUILD A
BUILDING, BRIDGE,
ROAD, TUNNEL,
DAM, OR TOWER

COMPARISON

IS
STRUCTURE
OK?

**CONSTRUCTION
PROCESSES**

OUTPUT
ACTUAL
STRUCTURE

BRIDGE, BUILDING,
DAM, ROAD,
TUNNEL, ETC.

MONITOR
INSPECTION
OF
CONSTRUCTION
PROJECT

FEEDBACK

OTHER OUTPUTS
(DANGEROUS CONDITIONS,
NOISE, ETC.)

134

PEOPLE IN CONSTRUCTION

1. Land owner

2. Architects and engineers

3. Civil engineers

4. General contractors

5. Estimators

6. Project managers

7. Tradespeople

PEOPLE IN THE CONSTRUCTION INDUSTRY

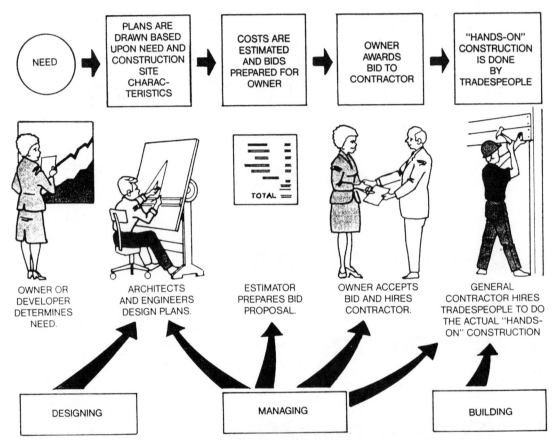

| NEED | → | PLANS ARE DRAWN BASED UPON NEED AND CONSTRUCTION SITE CHARAC-TERISTICS | → | COSTS ARE ESTIMATED AND BIDS PREPARED FOR OWNER | → | OWNER AWARDS BID TO CONTRACTOR | → | "HANDS-ON" CONSTRUCTION IS DONE BY TRADESPEOPLE |

OWNER OR DEVELOPER DETERMINES NEED.

ARCHITECTS AND ENGINEERS DESIGN PLANS.

ESTIMATOR PREPARES BID PROPOSAL.

OWNER ACCEPTS BID AND HIRES CONTRACTOR.

GENERAL CONTRACTOR HIRES TRADESPEOPLE TO DO THE ACTUAL "HANDS-ON" CONSTRUCTION

DESIGNING

MANAGING

BUILDING

THE THREE SUBSYSTEMS OF CONSTRUCTION (DESIGNING, MANAGING, AND BUILDING) INVOLVE CONTINUOUS COOPERATION AMONG THE OWNER, THE ARCHITECTS AND ENGINEERS, THE CONTRACTOR, AND THE TRADESPEOPLE.

TYPES OF STRUCTURES

BRIDGES

DAMS

ROADS

PIPELINES

AIRPORTS

BUILDINGS

HARBORS

TOWERS

CANALS

TUNNELS

TYPES OF BRIDGES

BEAM

CANTILEVER

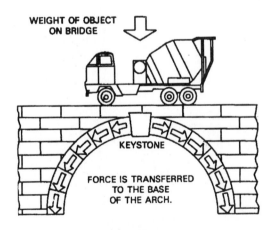

ARCH

SUSPENSION

138

TYPES OF BUILDINGS

RESIDENTIAL
(Hotels and housing)

COMMERCIAL
(Banks, stores, and offices)

INSTITUTIONAL
(Schools and hospitals)

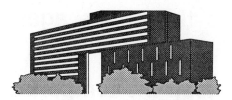

INDUSTRIAL
(Factories)

BUILDING A STRUCTURE

MAJOR CONCEPTS

After reading this chapter, you will know that:

■ Buildings are constructed in steps. The footing, foundation, floors, walls, and roof are built. Then the utilities and insulation are installed. Finally, the structure is finished.

■ The footing is the base of the foundation. It spreads the weight of the structure over a wider area of ground.

■ The foundation walls support the whole weight of the structure and transmit it to the footing.

■ Walls transmit the load from above to the foundation. They also serve as partitions between rooms.

■ The roof protects the house against weather and prevents heat loss.

■ Insulation helps keep the temperature of the house constant.

■ Utilities include plumbing, electrical, and heating systems.

■ Manufactured houses are built in a factory and taken to the construction site.

■ Wind effects must be taken into account in designing and building a skyscraper.

CHAPTER 11

BUILDING A STRUCTURE

KEY WORDS

Asphalt	Mortar
Bottom plate	National Electrical Code
Building permit	Pitch
Circuit breaker	Plumb
Floor joist	Plumbing
Footing	Rafter
Foundation wall	Sheathing
Framing	Sill plate
Frost line	Stud
Girder	Subfloor
Header	Top plate
Insulation	Truss
Manufactured housing	Vapor barrier
Mason	Wind load

THE CONSTRUCTION PROCESS INVOLVES

1. Choosing and preparing the site.

2. Building the foundation.

3. Building superstructure.

4. Installing utilities.

5. Finishing and enclosing the outside surfaces.

FOUNDATION

Usually made of concrete and supports the weight of the structure

Foundation (Substructure) Includes:

1. **EARTH.** . .on which the entire structure will rest

2. **FOOTING.** . .transfers the structure's weight to earth

3. **VERTICAL SUPPORTS.** . . which rest on the footing

SUPERSTRUCTURE

Is usually the part of the structure that is visible above the ground

Three Types Include:

1. Mass superstructures

2. Bearing wall superstructures

3. Framed superstructures

FOOTING

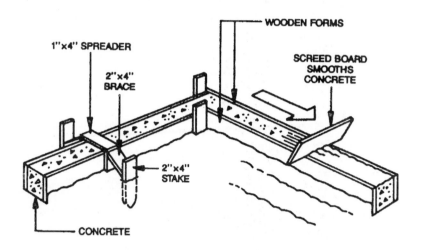

1"x4" SPREADER

2"x4" BRACE

WOODEN FORMS

SCREED BOARD SMOOTHS CONCRETE

2"x4" STAKE

CONCRETE

FOUNDATION WALL

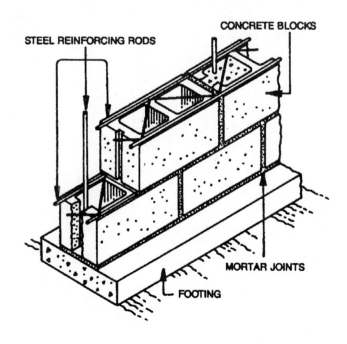

STEEL REINFORCING RODS

CONCRETE BLOCKS

MORTAR JOINTS

FOOTING

CONSTRUCTION OF A FLOOR

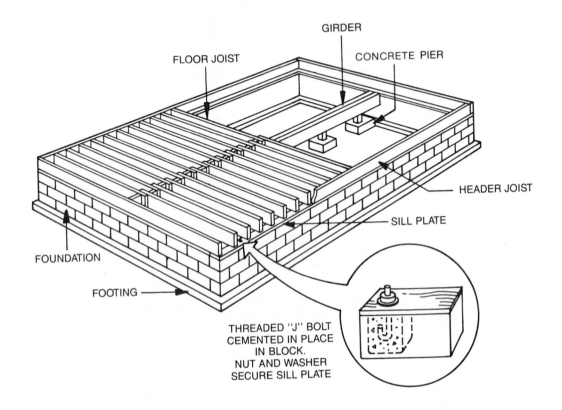

GIRDER

FLOOR JOIST

CONCRETE PIER

HEADER JOIST

SILL PLATE

FOUNDATION

FOOTING

THREADED "J" BOLT
CEMENTED IN PLACE
IN BLOCK.
NUT AND WASHER
SECURE SILL PLATE

FRAMING

ROOF TRUSSES

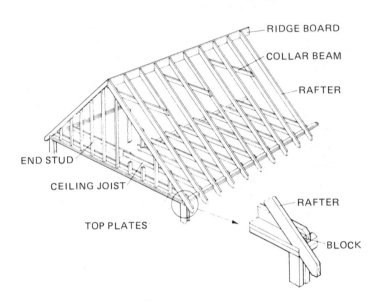

- RIDGE BOARD
- COLLAR BEAM
- RAFTER
- END STUD
- CEILING JOIST
- TOP PLATES
- RAFTER
- BLOCK

EXTERIOR WALL

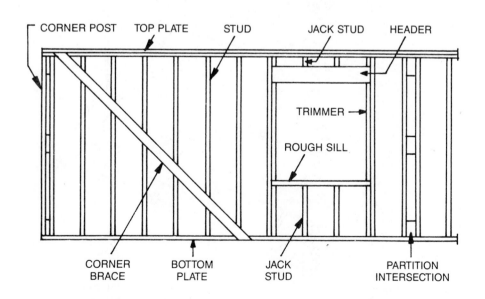

- CORNER POST
- TOP PLATE
- STUD
- JACK STUD
- HEADER
- TRIMMER
- ROUGH SILL
- CORNER BRACE
- BOTTOM PLATE
- JACK STUD
- PARTITION INTERSECTION

SPACING
BETWEEN STUDS

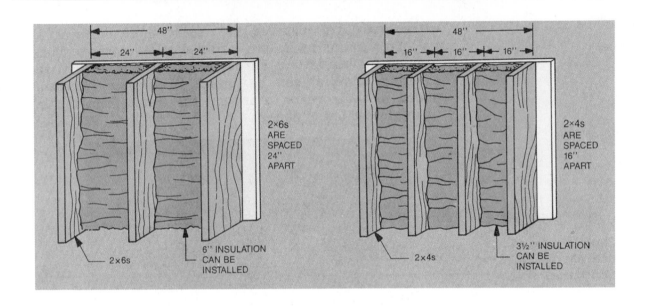

2×6s
ARE
SPACED
24"
APART

2×6s

6" INSULATION
CAN BE
INSTALLED

2×4s
ARE
SPACED
16"
APART

2×4s

3½" INSULATION
CAN BE
INSTALLED

ASPHALT SHINGLES

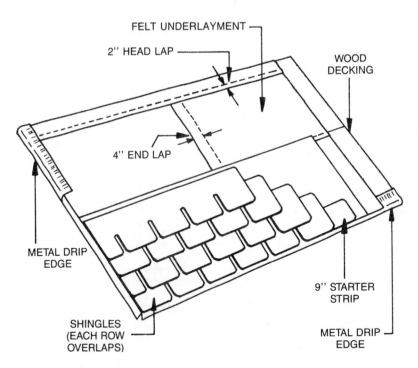

FELT UNDERLAYMENT

2" HEAD LAP

WOOD
DECKING

4" END LAP

METAL DRIP
EDGE

9" STARTER
STRIP

METAL DRIP
EDGE

SHINGLES
(EACH ROW
OVERLAPS)

PRINCIPAL TYPES
OF ROOFS

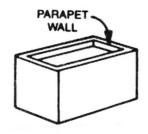

FLAT ROOF

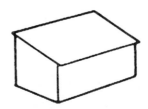

SHED ROOF

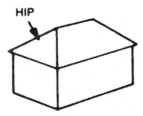

HIP ROOF

GABLE ROOF

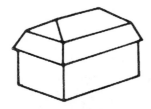

MANSARD ROOF

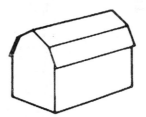

GAMBREL ROOF

DECK ROOF

PLUMBING SYSTEM IN A BATHROOM

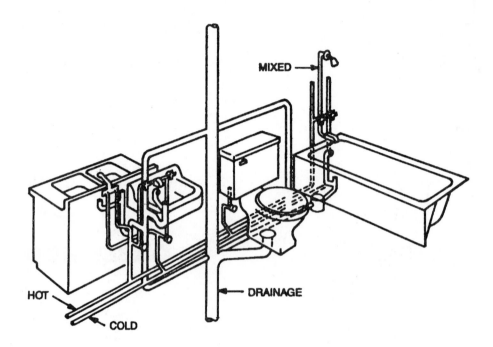

150

SERIES CIRCUIT

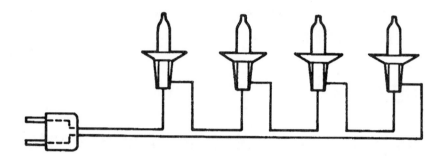

PARALLEL CIRCUIT

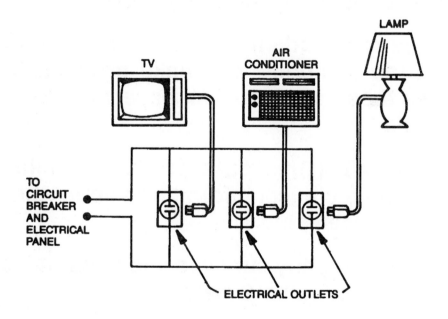

LAMP

TV

AIR
CONDITIONER

TO
CIRCUIT
BREAKER
AND
ELECTRICAL
PANEL

ELECTRICAL OUTLETS

CHAPTER 12

MANAGING PRODUCTION SYSTEMS

MAJOR CONCEPTS

After reading this chapter, you will know that:

■ A company's management must coordinate the work of different departments to make the best use of resources.

■ Architects design a building's shape and choose materials for it.

■ Engineers design a building's structure and its major systems.

■ A general contractor directs the work of many different people on a building project.

■ A company's marketing department decides what market the company will make products for, what features those products will have, and how they are to be sold.

■ A company that makes a product should provide service and technical information about the product after it has been sold.

■ Whether a company survives and does well depends on its financial management.

■ Managing includes planning, organizing, leading, and controlling.

MANAGING PRODUCTION SYSTEMS

KEY WORDS

Action item

Architect

Business plan

Cash flow analysis

Certificate of occupancy

Commission

Contract

Coordination

Dealer

Direct sales

Distributor

Engineer

EPA

Gantt chart

General contractor

Loss

Marketing

OSHA

Overruns

Owner

Permits

PERT chart

Productizing

Profit

Project manager

Prototype

Quality control

Research and development

Sales representative

Shareholder

Subcontractor

Throwaway

Union

Venture capitalist

Zoning

MANAGING PRODUCTION SYSTEMS

MANAGEMENT
Coordinates jobs of different departments

ARCHITECTS
Design building's shape and choose basic materials for it

ENGINEERS
Design building's structure and its major systems

GENERAL CONTRACTOR
Hire subcontractors and direct work of people on the construction job

MARKETING DEPARTMENT
Decides what market products are developed for, features of product, and how to sell the product

RESEARCH & DEVELOPMENT
Looks for new materials, process, and methods and applies these discoveries with previously used methods

154

SCHEDULING AND PROJECT TRACKING

Tools management can use to set and stay on schedules include:

GANTT CHART — bar graph that shows when work is to start and when it is to be finished.

PERT CHART — Program Evaluation Review Technique lists each step necessary to complete a project and the order and time needed.

These charts enable a manager to keep track of what is happening and move resources to most important jobs to stay on schedule.

SPREAD SHEET

SHOWS THE CASH FLOW OF A BUSINESS

SPREAD SHEET: SIX MONTH CASH FLOW						
	JAN	FEB	MAR	APR	MAY	JUNE
INCOME, DOLLARS	2400	3000	4500	4000	5000	5500
EXPENSES						
Rent	850	850	850	850	850	850
Telephone	116	173	164	175	204	227
Salaries	575	1270	2350	2350	2525	2610
Materials	255	478	763	652	940	985
Shipping	18	22	35	25	38	43
Insurance	125	125	125	125	125	125
TOTAL	1939	2918	4287	4177	4682	4840
INCOME−EXPENSES	461	82	213	−177	318	660
CASH TO START						
4000						
CASH IN BANK	4461	4543	4756	4579	4897	5557

GANTT CHART

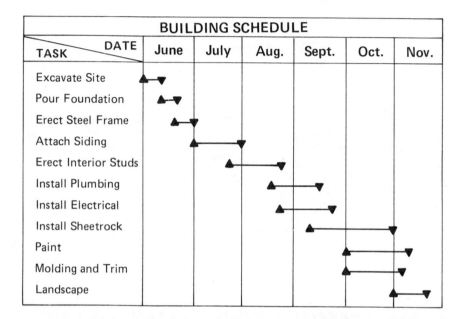

BUILDING SCHEDULE						
TASK \ DATE	June	July	Aug.	Sept.	Oct.	Nov.
Excavate Site	▲▼					
Pour Foundation	▲—▼					
Erect Steel Frame	▲—▼					
Attach Siding		▲———▼				
Erect Interior Studs		▲——————▼				
Install Plumbing			▲———▼			
Install Electrical			▲————▼			
Install Sheetrock				▲————▼		
Paint					▲———▼	
Molding and Trim					▲———▼	
Landscape						▲—▼

HELPS MANAGERS SCHEDULE AND MONITOR A PROJECT

PERT CHART

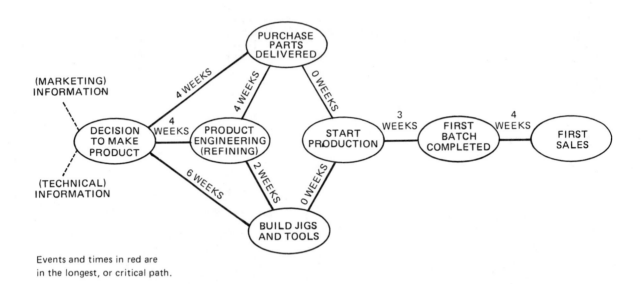

PURCHASE PARTS DELIVERED

(MARKETING) INFORMATION

4 WEEKS

4 WEEKS

0 WEEKS

DECISION TO MAKE PRODUCT

4 WEEKS

PRODUCT ENGINEERING (REFINING)

START PRODUCTION

3 WEEKS

FIRST BATCH COMPLETED

4 WEEKS

FIRST SALES

(TECHNICAL) INFORMATION

6 WEEKS

2 WEEKS

0 WEEKS

BUILD JIGS AND TOOLS

Events and times in red are
in the longest, or critical path.

MANAGEMENT TOOL THAT HELPS IN SCHEDULING AND MONITORING LARGE COMPLEX PROJECTS

ELEMENTS OF GOOD MANAGEMENT

Elements of good management in CONSTRUCTION and examples are:

PLANNING
1. Building design
2. Making financial arrangements

ORGANIZING
1. Choosing general contractor
2. Choosing subcontractors

LEADERSHIP
1. Promoting safety and good work habits
2. Resolving disputes quickly

CONTROL
1. Progress meetings

ELEMENTS OF GOOD MANAGEMENT

Elements of good management in MANUFACTURING and examples are:

PLANNING
1. Market research
2. Research and development

ORGANIZING
1. Making sure right people, parts, and machines are:
2. In the right place at the right time

LEADERSHIP
1. Selecting correct market
2. Making everyone interested in high quality

CONTROL
1. Quality control
2. Monitoring product sales

CHAPTER 13

ENERGY

MAJOR CONCEPTS

After reading this chapter, you will know that:

■ Work done on an object is equal to the distance it moves multiplied by the force used in the direction of the motion.

■ Energy is the ability to do work. It is the source of the force that is needed to work.

■ Kinetic energy is the energy of a moving object. Potential energy is the energy an object has because of its position, shape, or other features.

■ Potential energy can be changed into kinetic energy and kinetic energy can be changed into potential energy.

■ The principle of conservation of energy states that energy cannot be created or destroyed; it can only be changed from one form to another.

■ Energy sources are limited, unlimited, or renewable.

■ Most of the energy used in the United States today comes from limited energy sources.

ENERGY

KEY WORDS

Active solar	Hydroelectricity
Biomass	Kinetic energy
Conservation of energy	Nuclear fuel
Energy	Parabolic reflector
Fermentation	Passive solar
Fission	Potential energy
Fossil fuel	Quad
Fusion	Radiation
Gasohol	Solar Cell
Gasification	Work

WORK

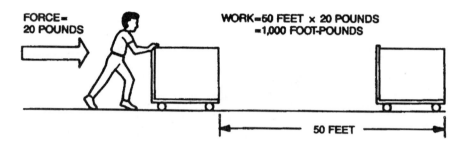

FORCE = 20 POUNDS

WORK = 50 FEET × 20 POUNDS = 1,000 FOOT-POUNDS

50 FEET

WORK. . .is equal to the distance an object moves, multiplied by the force in the direction of the motion.

ENERGY

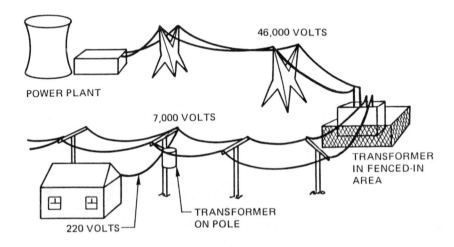

POWER PLANT

46,000 VOLTS

7,000 VOLTS

TRANSFORMER
IN FENCED-IN
AREA

TRANSFORMER
ON POLE

220 VOLTS

ENERGY. . .is the ability or capacity for doing work. It is the source of the force that is needed to do work.

KINETIC ENERGY

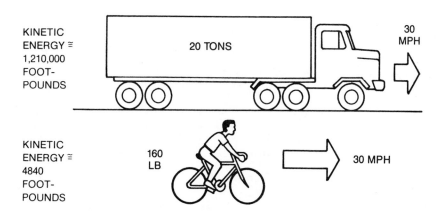

KINETIC ENERGY ≅ 1,210,000 FOOT-POUNDS

20 TONS

30 MPH

KINETIC ENERGY ≅ 4840 FOOT-POUNDS

160 LB

30 MPH

KINETIC ENERGY. . .is energy in motion.

POTENTIAL ENERGY

POTENTIAL ENERGY
STORED IN SPRING

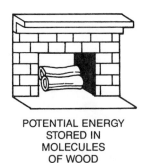

POTENTIAL ENERGY
STORED IN
MOLECULES
OF WOOD

POTENTIAL ENERGY
STORED IN POSITION
ABOVE FLOOR

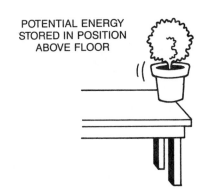

POTENTIAL ENERGY. . .is energy that is stored in an object due to its position, shape, or other feature.

CLASSIFICATIONS OF ENERGY SOURCES

LIMITED
Coal
Oil
Natural Gas
Uranium

UNLIMITED
Solar
Wind
Gravitational
Tidal
Geothermal
Fusion

RENEWABLE
Wood
Biomass Gasification
Biomass Fermentation
Animal Power
Human Muscle Power

ENERGY IN THE UNITED STATES

SOURCES
Sources of Energy in the United States

SOURCES OF ENERGY IN
THE UNITED STATES

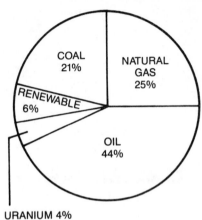

COAL
21%

NATURAL
GAS
25%

RENEWABLE
6%

OIL
44%

URANIUM 4%

USES
Uses of Energy in the United States

USES OF ENERGY IN
THE UNITED STATES

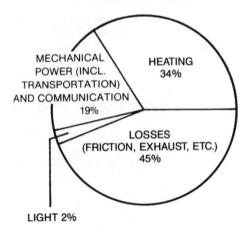

MECHANICAL
POWER (INCL.
TRANSPORTATION)
AND COMMUNICATION
19%

HEATING
34%

LOSSES
(FRICTION, EXHAUST, ETC.)
45%

LIGHT 2%

THE FORMATION OF COAL

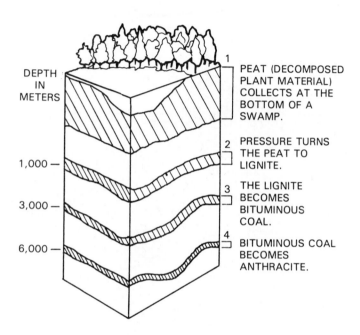

DEPTH
IN
METERS

1,000 —

3,000 —

6,000 —

1   PEAT (DECOMPOSED
PLANT MATERIAL)
COLLECTS AT THE
BOTTOM OF A
SWAMP.

2   PRESSURE TURNS
THE PEAT TO
LIGNITE.

3   THE LIGNITE
BECOMES
BITUMINOUS
COAL.

4   BITUMINOUS COAL
BECOMES
ANTHRACITE.

THE MAKING OF NUCLEAR ENERGY

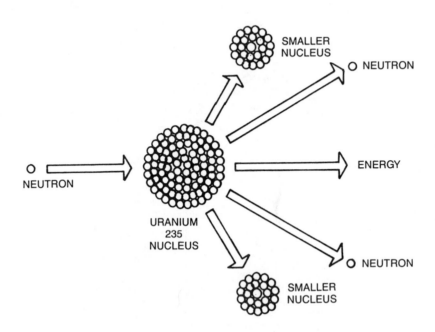

When a large atomic nucleus is split, a very small amount of its matter is converted into an enormous amount of energy.

THE MAKING OF HYDROELECTRICITY

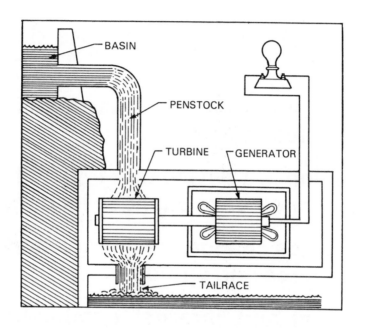

Water falling through a penstock turns the blades of a turbine. The turbine is connected to a generator, which produces electricity.

POWER

MAJOR CONCEPTS

After reading this chapter, you will know that:

■ Power is the amount of work done during a given period of time.

■ An engine is a machine that uses energy to create mechanical force and motion.

■ A transmission is a device that transmits force from one place to another or changes its direction.

■ Modern engines change the energy stored in fuel to mechanical force and motion.

■ Both external and internal combustion engines change the potential energy stored in fuel into heat. The heat expands a gas, which moves a piston.

■ Newton's third law of motion states that for every action, there is an equal and opposite reaction.

■ A generator changes rotary motion into electrical energy.

■ An electric motor changes electrical energy into rotary motion.

CHAPTER 14

POWER

KEY WORDS

Absolute zero

Alternating current

Battery

Diesel engine

Electric motor

Engine

External combustion engine

Four-stroke cylinder

Fractional horsepower

Generator

Horsepower

Hydraulic

Idler wheel

Internal combustion engine

Jet engine

Load

Momentum

Nuclear reactor

Pneumatic

Poles

Power

Power system

Pressure

Reaction engine

Rocket engine

Rotor

Stator

Superconductor

Transformer

Transmission

Two-stroke cycle

Watt

ENGINES

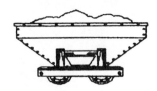

Different kinds of engines use different kinds of fuels. How are these used?

EXTERNAL COMBUSTION ENGINES
(Burns fuel such as coal or wood)

INTERNAL COMBUSTION ENGINES
(Uses a fast burning fuel - gas or diesel)

REACTION ENGINES
(Use jet fuel and air and burns rapidly)

ELECTRIC MOTORS
(Use a power source such as a battery)

NUCLEAR REACTORS
(Uses heat from fission of uranium-235 to boil liquid to form steam to run turbines)

EXTERNAL COMBUSTION ENGINE

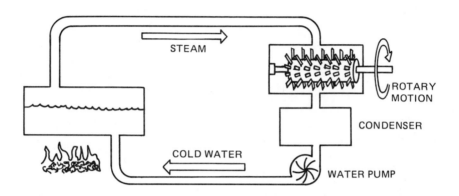

The force of the expanding steam is converted into rotary motion by the turbine.

INTERNAL COMBUSTION ENGINE

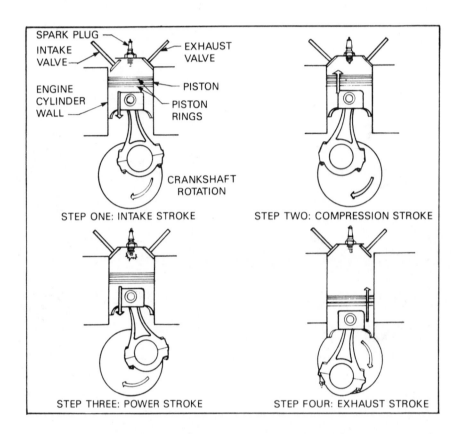

SPARK PLUG
INTAKE VALVE
EXHAUST VALVE
ENGINE CYLINDER WALL
PISTON
PISTON RINGS
CRANKSHAFT ROTATION

STEP ONE: INTAKE STROKE

STEP TWO: COMPRESSION STROKE

STEP THREE: POWER STROKE

STEP FOUR: EXHAUST STROKE

AUTOMOBILE ENGINE

REACTION
ENGINE

A JET ENGINE IS A GOOD EXAMPLE OF NEWTON'S THIRD LAW:

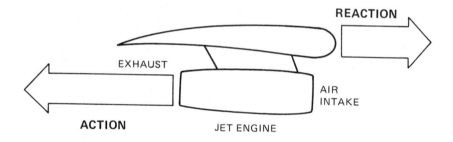

FOR EVERY ACTION THERE IS AN EQUAL AND OPPOSITE REACTION

NUCLEAR POWER

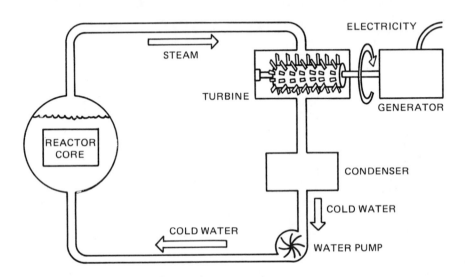

ELECTRICITY

STEAM

TURBINE

GENERATOR

REACTOR CORE

CONDENSER

COLD WATER

COLD WATER

WATER PUMP

Steam turns a turbine that is connected to a generator that converts the rotary mechanical motion to electrical energy.

TRANSMISSIONS

. . .A transmission transfers, or carries, force from one place to another or changes its direction.

MECHANICAL

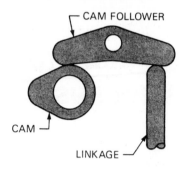

CAM FOLLOWER

CAM

LINKAGE

PNEUMATIC

HYDRAULIC

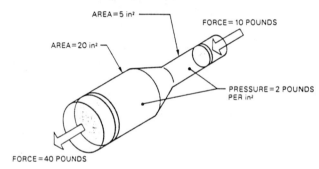

AREA = 5 in²

FORCE = 10 POUNDS

AREA = 20 in²

PRESSURE = 2 POUNDS PER in²

FORCE = 40 POUNDS

MAGNETIC

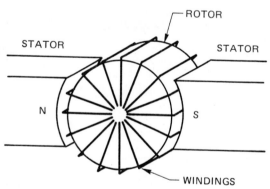

ROTOR

STATOR

STATOR

N

S

WINDINGS

ELECTRICAL

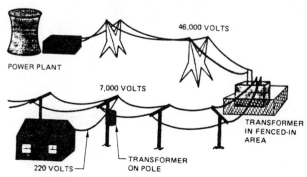

46,000 VOLTS

POWER PLANT

7,000 VOLTS

TRANSFORMER IN FENCED-IN AREA

220 VOLTS

TRANSFORMER ON POLE

BATTERIES

Batteries are used to store energy in a power system.

Cars use batteries with alternators to recharge the battery as it is being used.

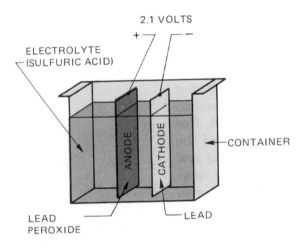

SIMPLE LEAD-ACID BATTERY CELL

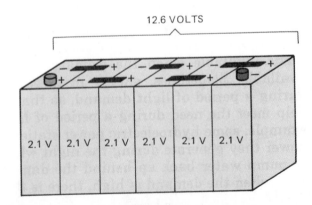

TYPICAL 12-VOLT CAR BATTERY

TRANSPORTATION

MAJOR CONCEPTS

After reading this chapter, you will know that:

■ A transportation system is used to move people or goods from one location to another.

■ Modern transportation systems have helped to make countries interdependent.

■ The availability of rapid, efficient transportation systems has changed the way we live.

■ Transportation systems convert energy into motion.

■ Steam was the first important source of mechanical power for transportation systems.

■ Modern transportation systems often use internal combustion engines or electric motors.

■ Intermodal transportation systems make optimum use of each type of transportation used in the system.

■ Most transportation systems use vehicles to carry people or goods, but some systems do not use any vehicles.

KEY WORDS

Buoyancy	Piggyback
Commute	Pipeline
Container ship	Propeller
Conveyor	Rocket
Diesel	Steam engine
Drag	Thrust
Engine	Transmission
Intermodal	Turbo-prop
Internal combustion engine	Vehicle
Jet	Weight
Lift	

RESOURCES FOR TRANSPORTATION SYSTEMS

TOOLS & MACHINES
(Vehicles & Pipelines)

PEOPLE
(Pilots)

ENERGY
(Gasoline)

INFORMATION
(Computers)

MATERIALS
(Composites)

CAPITAL
(Money)

TIME
(Hours)

SYSTEMS DIAGRAM
TRANSPORTATION SYSTEMS

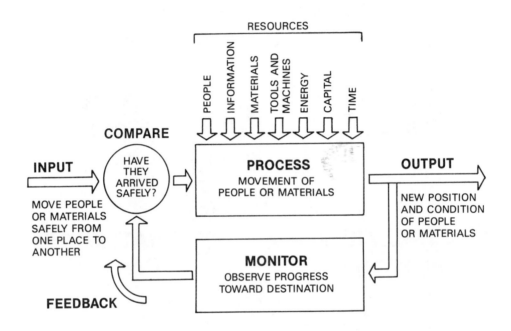

Like all other systems, transportation systems can be represented with system diagrams.

CLASSIFICATIONS OF TRANSPORTATION SYSTEMS

NONVEHICLE

LAND

AIR

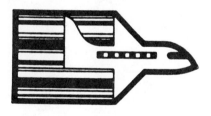

SEA

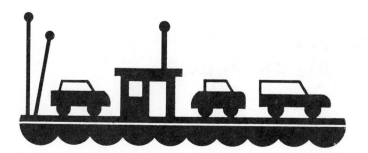

SPACE

HOW PLANES FLY

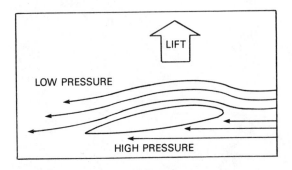

AIRFLOW OVER AN AIRPLANE WING

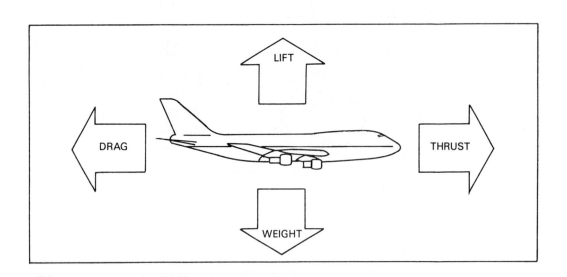

THE FOUR FORCES
ACTING ON AN AIRPLANE

WHAT MAKES A BOAT FLOAT?

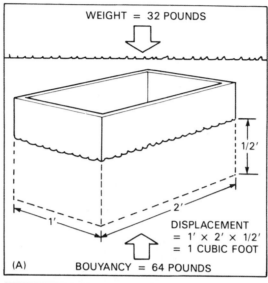

WEIGHT = 32 POUNDS

1/2'

2'

1'

DISPLACEMENT
= 1' × 2' × 1/2'
= 1 CUBIC FOOT

(A) BOUYANCY = 64 POUNDS

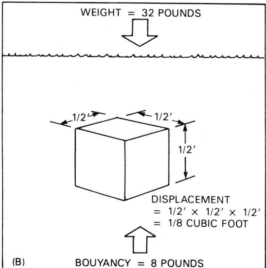

WEIGHT = 32 POUNDS

1/2' 1/2'

1/2'

DISPLACEMENT
= 1/2' × 1/2' × 1/2'
= 1/8 CUBIC FOOT

(B) BOUYANCY = 8 POUNDS

Buoyancy equals the weight of the water that is displaced by an object. (A) A box floats in water. (B) A solid piece of metal weighing the same as the box sinks in water.

FOUR KINDS OF AIRCRAFT ENGINES

INTERNAL COMBUSTION ENGINE & PROPELLER

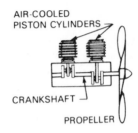

JET ENGINE

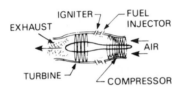

TURBO-PROP ENGINE

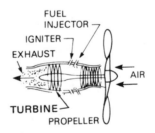

ROCKET ENGINE

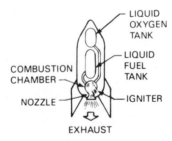

EXTERNAL COMBUSTION ENGINE (STEAM)

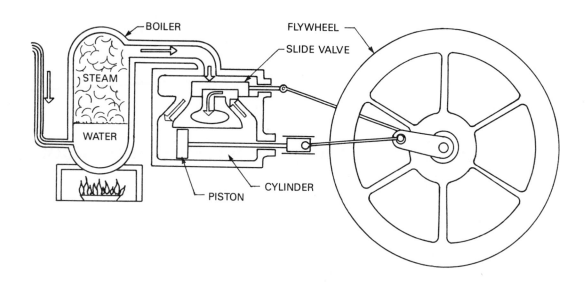

Steam engines are external combustion engines. The fuel is burned in an open chamber. Steam engines burn wood, coal, oil, or other fuel to heat a boiler containing water, creating steam. The steam pressure is used to push a piston. The piston's reciprocating (back-and-forth) motion is converted into rotary motion. The rotary motion can then be transferred to wheels. Steam locomotives need to stop frequently to take on new supplies of water and fuel, which they must carry with them.

CHAPTER 16

IMPACTS FOR TODAY AND TOMORROW

MAJOR CONCEPTS

After reading this chapter, you will know that:

■ Outputs of a technological system can be desired, undesired, expected, or unexpected.

■ People determine whether technology is good or bad by the way they use it.

■ Technology produces many positive outputs and solves many problems. Sometimes, however, negative outputs create new problems.

■ Technology must be fitted to human needs.

■ Technology must be adapted to the environment.

■ Existing technological systems will act together to produce new, more powerful technologies.

■ Using futuring techniques, people can anticipate the consequences of a new technology.

IMPACTS FOR TODAY AND TOMORROW

KEY WORDS

Acid rain	Intelligent buildings
Confluence	Magnetic levitation (Maglev)
Environment	Manufacturing in space
Ergonomics	Personal rapid transit (PRT)
Fusion	Pollution
Futures wheel	Smart houses
Futuring	Speech synthesis
Growth hormone	Sulfur dioxide
Halophyte	Videoconference
Impact	Voice recognition

FOUR KINDS OF OUTPUTS OF TECHNOLOGY

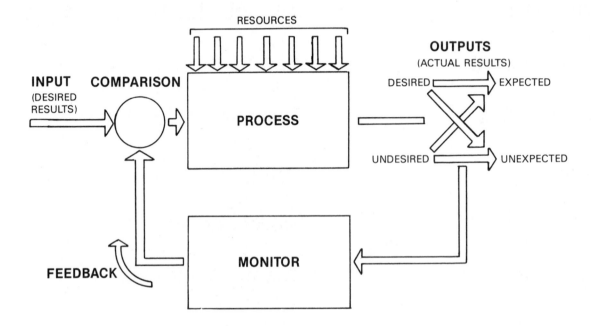

RESOURCES

OUTPUTS
(ACTUAL RESULTS)

INPUT
(DESIRED RESULTS)

COMPARISON

PROCESS

DESIRED — EXPECTED

UNDESIRED — UNEXPECTED

FEEDBACK

MONITOR

Technological systems create four possible kinds of outputs.

SYSTEM DIAGRAM
FUTURE TECHNOLOGIES

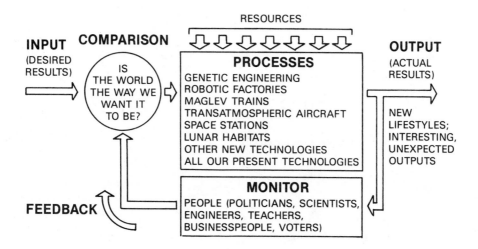

This shows how future
technologies will affect our lives.

CONFLUENCE OF SYSTEMS OF MODERN TECHNOLOGY

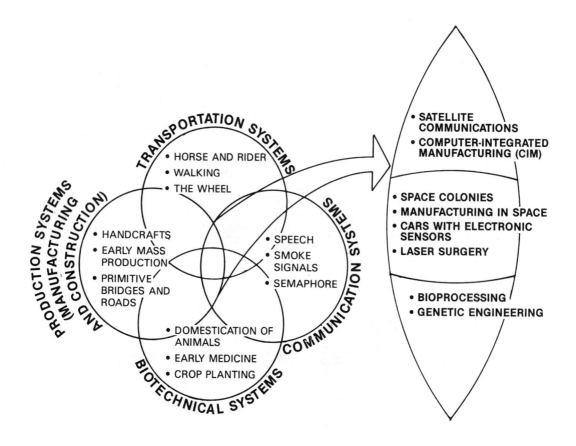

TRANSPORTATION SYSTEMS

PRODUCTION SYSTEMS (MANUFACTURING AND CONSTRUCTION)

COMMUNICATION SYSTEMS

BIOTECHNICAL SYSTEMS

- HORSE AND RIDER
- WALKING
- THE WHEEL

- HANDCRAFTS
- EARLY MASS PRODUCTION
- PRIMITIVE BRIDGES AND ROADS

- SPEECH
- SMOKE SIGNALS
- SEMAPHORE

- DOMESTICATION OF ANIMALS
- EARLY MEDICINE
- CROP PLANTING

- SATELLITE COMMUNICATIONS
- COMPUTER-INTEGRATED MANUFACTURING (CIM)

- SPACE COLONIES
- MANUFACTURING IN SPACE
- CARS WITH ELECTRONIC SENSORS
- LASER SURGERY

- BIOPROCESSING
- GENETIC ENGINEERING

FUTURING TECHNIQUES
FUTURES WHEEL

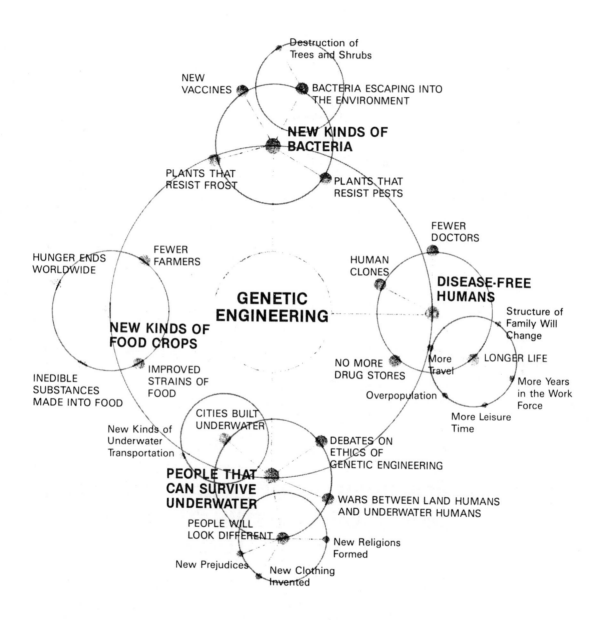

A futures wheel helps us determine possible outcomes of a new idea or technology. In this case, possible outcomes of engineering are forecast.

FUTURING TECHNIQUES
CROSS-IMPACT ANALYSIS

	GENETICALLY ENGINEERED HUMANS	CITIES ON THE MOON	LIFE ON OTHER PLANETS	CITIES UNDER THE SEA
AIR POLLUTION	• Humans will be able to breathe sulphur dioxide	• Moon societies will safeguard against the kind of pollution that occurred on earth	• People will vacation on nonpolluted planets	• Polluted air will be used as fuel by sea cities
RISING ENERGY COSTS	• Humans will be able to survive on fewer calories	• Energy will be beamed to earth from a moon base	• Energy sources will be exported from other planets to earth	• There will be new uses of sea water to provide energy
OVERPOPULATION	• Child-bearing will be forbidden • Only genetically engineered humans needed for specific purposes will be authorized • People will debate whether we should alter human life	• Societies will move to the moon	• Marriages will occur between earthlings and space beings • Attempts to transfer populations to other planets will meet with resistance; interspace wars will result	• People will work in new industries like mining sea beds
LONGER LIFE SPANS	• People will be altered several times during their lifetimes as conditions change on earth	• People will commute to "summer homes" on the moon • New travel agencies will specialize in moon-earth travel	• People will spend more time traveling to distant planets	• People will live part of their lives on land and part in sea cities to learn more about all forms of life

SEVERAL POSSIBLE FUTURES ARE IDENTIFIED.
IMPACT OF A COMBINATION
OF FUTURES IS STUDIED.

APPENDIX B CHAPTER REVIEW QUESTIONS AND ANSWERS

CHAPTER 1 REVIEW QUESTIONS AND ANSWERS

1. Explain the difference between science and technology.
   Answer: Science is the study of why natural things happen the way they do. Technology is the use of knowledge by people to turn resources into goods and services that society needs.

2. Describe how technology affected your routine this morning.
   Answer: Might relate to cooking breakfast, washing and grooming, transportation to school.

3. List five needs that people have and explain how technology helps to satisfy those needs.
   Answer: Need for food (agriculture, freeze drying, microwave ovens); Need for medical care (medicine, medical tools); Need for consumer items (manufactured products); Need to produce energy (utilities); Need to communicate (satellite, telephone computer); Need for transportation (vehicles, conveyors, pipelines).

4. Explain two ways in which technology satisfies our need to communicate ideas or process information.
   Answer: Devices have been produced to facilitate electronic and graphic communications (cameras, pencils, radios, televisions, satellites, computers, etc.)

5. Give two examples of technologies that have developed from more simple technologies.
   Answer: Steamship (shipbuilding techniques and the steam engine); Photography (optics and chemistry); Newspaper (telegraph and printing technology); Automobile (internal combustion engine and wagon carriage construction technology); Artificial heart (synthetic materials and medical technology); Airplane (lightweight gasoline engine and kite/glider technology); Spy aircraft (new materials (titanium) and jet engine technology); Exploration of the earth (photographic technology and satellite technology)

6. Draw a technological time line that illustrates how technology is growing at an exponential rate.
   Answer: See page 17, and Technological Time Line on pages 494–497.

7. Describe one major example of how technology has made life easier for people.
   Answer: Might relate to any area that makes one's routine easier, or any of the areas in which technology helps to satisfy our needs (agricultural, medical, manufacturing, energy, communication, and transportation)

8. Define agricultural era, industrial era, and information age.
   Answer: In the agricultural era (about 8000 B.C. to A.D. 1750), most people lived by raising crops and animals and processing them into food, clothing, and other products. In the industrial era (beginning about 1750, the Industrial Revolution precipitated the invention of a great many mechanical devices), mechanical power replaced human and animal power. In the information age (the present era), much of technology is based upon electronics and the computer. Computer control is replacing human control in many operations. Most people today are engaged in processing information, and produce knowledge or new information as a part of their work.

CHAPTER 2 REVIEW QUESTIONS AND ANSWERS

1. What are the seven resources common to all technologies?
   Answer: People, information, materials, tools and machines, energy, capital, and time.

2. Define a machine in your own words. Then use your definition to determine whether the following items are machines:
   a. a baseball bat
   b. software for a video game
   c. a radio
   d. a ramp for wheelchairs
   e. a hand-operated drill
   f. a wrench

   Answer: Machines change the amount, speed, or direction of a force. By that definition, even a simple device like an inclined plane can be considered to be a machine. Permit the student some license in the definition. A rigorous definition cannot be expected from the junior high/middle school student. What should be interesting is whether student will rewrite the definition to make it more rigorous when forced to classify devices as machines when in fact he or she does not think that they are. For example, a baseball bat is a lever and by our definition, is a machine because it changes the speed and direction of a force. Will the students determine that machines must have moving parts? Should radios and computers be considered to be electronic machines since they involve the movement of electrons? Permit the class to decide.

3. Wood is a renewable resource. Does that mean that we can cut down all the trees we need? Explain your answer.
   Answer: Although wood is a renewable resource, its use must be carefully planned. Indiscriminate harvesting of forest resources can lead to erosion of arable land. Lumber companies plan their harvests systematically. Portions of their acreage are clear-cut in a checkerboard pattern. These areas are then re-seeded with rapidly growing varieties of trees and are given time to mature while other areas are cut.

4. What are two advantages of synthetic materials over natural materials?
   Answer: a. Synthetic materials can replace scarce natural materials. b. Synthetic materials can be made with improved properties. c. Synthetic materials can sometimes be less costly than natural materials. d. Sometimes, synthetic materials are easier to produce than natural material that may be difficult to obtain.

5. Name three of the limited and three of the unlimited sources of energy.
   Answer: Limited: a. Coal b. Oil c. Natural Gas d. Nuclear Fission
   Unlimited: a. Solar b. Wind c. Gravitational d. Tidal e. Geothermal f. Nuclear Fusion

6. If you wanted to start a company, how could you arrange to get capital?
   Answer: Sell stock to private investors, borrow money from banks, invest your own money.

7. How do knowledge and information differ?
   Answer: Information is processed data. Knowledge is information made useful by humans, and includes conclusions drawn by people, based upon available information.

CHAPTER 3 REVIEW QUESTIONS AND ANSWERS

1. Give one example each of problems involving society, the environment, and the individual.
   Answer: Examples can be taken from page 45. They are: disposing of wastes without harming the environment, (environmental problem), producing enough energy to meet our increasing needs (society problem), assuring a continuing supply of clean, safe water (environmental problem), improving the sound quality of recorded music (individual problem), helping vision-impaired people to see better (society problem), helping automobile drivers communicate with each other (individual problem).

2. Propose a workable and economical solution to a problem involving a personal issue.
   Answer: The problem and solution may be similar to the following, or teachers may create their own. Problem situation 1: Young children enjoy playing with toys that involve balance and dexterity. Design Brief 1: Design and construct a toy that will either balance on, or move along, a taut piece of string. The finished object should be attractive and safe for a small child to use.

3. What are the seven problem-solving steps listed in this chapter?
   Answer: 1. Describe the problem as clearly and fully as you can. 2. Describe the results you want. 3. Gather information. 4. Think of alternative solutions. 5. Choose the best solution. 6. Implement the solution you have chosen. 7. Evaluate the solution and make necessary changes.

4. What are five ways of coming up with alternative solutions?
   Answer: Past experience, brainstorming, trial and error, insight, and by accident.

5. Your city has run out of land for landfill (refuse disposal). City government has chosen to build a very expensive incinerator to handle the garbage problem. What are some trade-offs that were made in reaching this decision?
   Answer: Trade-offs that were made could include the following: Since there is no more land for landfills building the incinerator would mean the city would not have to look elsewhere for land; the incinerator, though expensive to build, might not require a high level of maintenance. A good class discussion or debate could happen over this question.

6. Give an example of how a person's values might affect his or her decision about the kind of car to buy.
   Answer: If a person thinks of an automobile only as transportation, he or she might decide to buy a basic car that gets good gas mileage. If a person feels that automobiles are neat and driving is fun, he or she might decide to buy a sports car or a luxury car.

7. List five important parts of a technological system.
   Answer: Input, process, output, monitor, feedback.

8. What does the process part of the system do?
   Answer: The process part of the system is the action part of the system. It combines the resources and produces results.

9. Why is feedback important in a system?
   Answer: Feedback is important in a system because it uses the information about the output(s) of the system in order to modify the system's process.

10. Give an example of:
    a. feedback you've received recently at school
    b. feedback you've received from a friend
    c. feedback you can receive when riding a bicycle
    d. feedback in a technological system
    Answer: a. Results from a test, quiz, or homework; comments about your performance from teachers or peers; b. Congratulations, an argument, or comments about your appearance; c. Information on the weather, road conditions, or the condition of your bicycle (are the tires flat?); d. Experimentation with different parts of the system will give feedback in the form of actual results.

11. Give an example of feedback in a body system.
    Answer: Our bodies maintain a temperature of about 98.6 degrees Fahrenheit. When our body temperature goes up or down due to weather changes or illness, we receive feedback in the form of sweating (to cool the body down) or shivering (to warm the body up).

12. Name some subsystems that make up a large railroad system.
    Answer: Vehicle system, communication system, management system, power system, lighting system, suspension system, drive train system, ignition system, lubrication system, fuel system.

13. Using the basic system model, model the operation of a nuclear power plant. Indicate the input (desired result), the process, the output (actual result), monitoring, and comparison.
Answer: See Figure below.

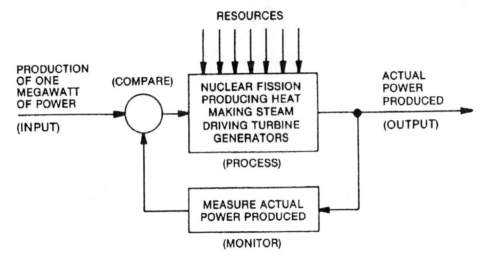

System diagram of a nuclear power plant

14. The automobile has become a very important system of transportation. Identify an output resulting from the development of the automobile that is:
a. expected and desirable.
b. expected and undesirable.
c. unexpected and desirable.
d. unexpected and undesirable.
Answer: a. People are able to move quickly and easily from one place to another. b. Accidents involving automobiles cause injuries and deaths. c. The widespread use of cars has created whole new industries that did not exist before, providing new jobs (e.g., car washes, motels, auto parts stores, car repair stores of all kinds). d. Automobiles are used in such large numbers that they cause serious air pollution in some parts of the country.

CHAPTER 4 REVIEW QUESTIONS AND ANSWERS

1. Name three particles found inside of atoms.
Answer: Protons, electrons, and neutrons.

2. In an electric circuit powered by a 9-volt battery, one ampere of current flows. If the 9-volt battery is replaced by a 20-volt battery, does more or less current flow? Why?
Answer: More current will flow because as voltage increases, current increases (Ohm's Law: I(amps) = E(volts)/R(ohms)). Note that you will not get one ampere for very long from a standard 9-volt battery.

3. Name five electronic components that can be used to build circuits.
Answer: Resistors, capacitors, inductors, transistors, diodes, batteries, switches, light-emitting diodes (LEDs), integrated circuits (many kinds), other semiconductors (SCRs, triacs), etc.

4. Why was the invention of the integrated circuit important in the history of technology?
Answer: The integrated circuit allows very complex functions to be built into tiny spaces. Combining many integrated circuits allows us to build very sophisticated equipment in a small space at low cost. Computers that used to cost hundreds of thousands of dollars and occupy large

rooms have been replaced by computers that can sit on top of a desk and are affordable for many consumers. Integrated circuits have not only made information processing tools available to the general public; they have also allowed the introduction of computers into almost all other technologies (appliances, cars, airplanes, clothing manufacturing, food preparation, etc.). The advent of the integrated circuit contributed to the rise of the information age.

5. Based on your own experience and observation, give one example each of how the use of electronics has changed manufacturing, transportation, communication, and health care technologies.
Answer: Individual evaluation. Examples that may be given are: in manufacturing, robots and numerically controlled machines routinely perform fabrication, welding, painting, and other operations. In transportation, electronics allow remote control of trains, rapid communication throughout the transportation system, and improved control and operation of the vehicles in use. Electronic computers and communications (satellite, fiber optic, radio, and wire) have revolutionized the information industry. In the health care field, electronics has provided us with new tools for diagnosing and treating illnesses, as well as making possible some artificial, operating body parts (arms, legs).

6. Should a telephone be an analog or a digital instrument? Why?
Answer: This same question is asked again at the end of Chapter 7. Having only read Chapter 4, the student is expected to answer that a telephone should be an analog instrument because it operates on a smoothly varying non-digital signal—the human voice. Some students may already know that there are advantages to be gained by converting this signal to a digital signal, but most will see that as a new concept in Chapter 7.

7. Name four parts of a computer.
Answer: Processor, Memory, Input, Output.

8. Describe the difference between operating system software and applications software.
Answer: Operating system software enables the user to control and access the computer's memories, printers, and other attached devices. Applications software provides instructions to the computer to perform specific, well-defined tasks.

9. A computer can be used for mailing letters to thousands of people. List the major components that you would expect to find in such a computer system. Draw the system using a general system model.

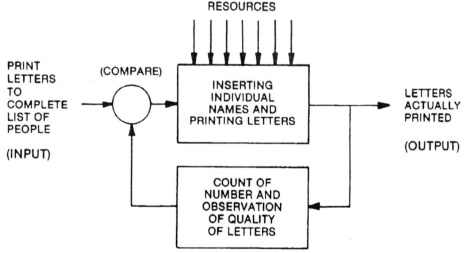

Answer: Such a computer would include a processor, a memory for storing the program, an input device of some kind, and a printer of some kind for an output device. The computer would have to have both operating system software and an applications program. The names and addresses to be used in the mailing would be entered through the input device, which might be magnetic tape or an optical character reader. The printer ideally would be capable of producing good quality ("letter quality") print at a rapid rate to complete the job in a timely fashion.

10. Do you think it is a good idea to have a totally automated system with no involvement by people? Why or why not? Give an example to support your answer.
    Answer: This question should encourage independent thinking and produce opposite (correct) points of view, depending upon the examples chosen.

CHAPTER 5 REVIEW QUESTIONS AND ANSWERS

1. What is communication? Define it in your own words. Then use your definition to tell which of the following are not communication, and why not.
   a. a baby crying for its mother
   b. the sound of glass being broken accidentally
   c. a car horn
   d. a railroad car screeching on the tracks
   e. a bird chirping
   f. static on the radio
   Answer: Communication includes having a message sent, received, and understood. By this definition, (a), (c), and (e) would be considered communication. A baby communicates to its mother by crying; a car horn is sounded as a signaling device and is meant to communicate the presence of a car; a bird calls to other birds by chirping. Sounds are not always communication. A sound made by breaking glass (b) might not be received or understood; a railroad car screeching on the tracks (d) is noise, as is (e) static on the radio.

2. Give an example of animal-to-human communication.
   Answer: A dog barking for a treat.

3. Give an example of machine-to-human communication and an example of human-to-machine communication.
   Answer: Machine-to-human — a clock; a traffic light; a smoke alarm. Human-to-machine — typing on a typewriter or word processor; speaking into a voice recognition computer; programming a microwave oven.

4. Choose a communication system you use and tell how it uses the seven technological resources.
   Answer: See system diagram of radio communication system on page 125.

5. What are four purposes of communication?
   Answer: a. to inform b. to persuade c. to educate d. to entertain

6. What are the three parts of the process of a communication system?
   Answer: The transmitter, the channel, and the receiver.

7. Give two examples of noise that you do not hear.
   Answer: a. Smudges on a drawing. b. "Snow" on a TV.

8. Draw a diagram of a person-to-person communication system. Label the transmitter, the channel, and the receiver.

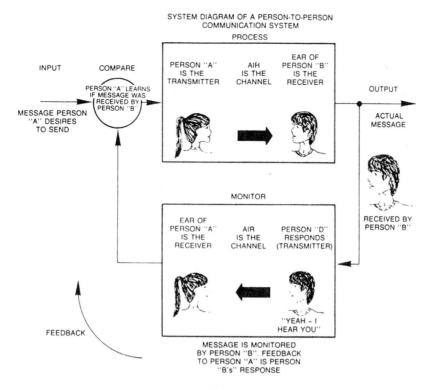

SYSTEM DIAGRAM OF A PERSON-TO-PERSON
COMMUNICATION SYSTEM

PROCESS

INPUT   COMPARE

PERSON "A" LEARNS
IF MESSAGE WAS
RECEIVED BY
PERSON "B"

MESSAGE PERSON
"A" DESIRES
TO SEND

PERSON "A"
IS THE
TRANSMITTER

AIR
IS THE
CHANNEL

EAR OF
PERSON "B"
IS THE
RECEIVER

OUTPUT

ACTUAL
MESSAGE

RECEIVED BY
PERSON "B"

MONITOR

EAR OF
PERSON "A"
IS THE
RECEIVER

AIR
IS THE
CHANNEL

PERSON "D"
RESPONDS
(TRANSMITTER)

"YEAH – I
HEAR YOU"

FEEDBACK

MESSAGE IS MONITORED
BY PERSON "B". FEEDBACK
TO PERSON "A" IS PERSON
"B's" RESPONSE

9. What kind of communication system would be best for each of the following?
   a. selling a used stereo
   b. asking for money for cancer research
   c. letting people know you are looking for a part-time job
   d. entertaining a large audience
   Answer: a. newspaper ads, advertise on bulletin board; b. flyers, local newspaper ads; c. local newspaper ad, notice on local bulletin board, word of mouth; d. public address system, slide projector, overhead projector, video monitor with large screen or projection device.

10. Describe the difference between graphic and electronic communication. Give one example of each kind of communication.
    Answer: In graphic communication, the channel carries images or printed words (writing, freehand drawing and sketching, technical drawing, photography, printing). In electronic communication systems, the channel carries electrical signals (telegraph, telephone, radio, television, microwave, satellite, data communication, facsimile, recording systems).

CHAPTER 6 REVIEW QUESTIONS AND ANSWERS

1. What is graphic communication?
   Answer: Graphic communication systems use images or printed words to convey a message.

2. Make an isometric and an orthographic drawing of a 2" x 2" x 1" block.
   Answer: See the following figure.

Answer to Question 2.

ISOMETRIC DRAWING OF 2" x 2" x 1" BLOCK

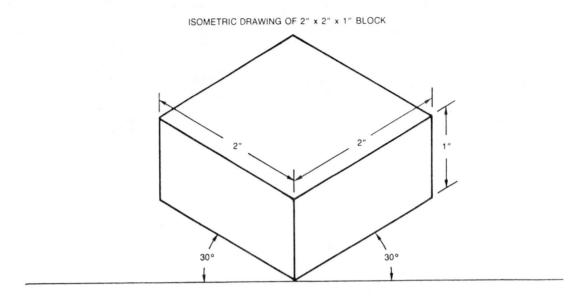

ORTHOGRAPHIC DRAWING OF 2" x 2" x 1" BLOCK

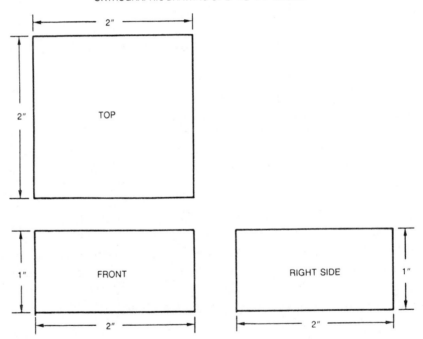

3. List two ways CAD has changed technical drawing.
   Answer: 1. Doing design and drafting together saves time. 2. CAD reduces the chance of error. 3. CAD makes it possible to save time on drawing. 4. Permits easier revision of drawings. 5. Improves accuracy and consistency.

4. Using the library as a resource, name five careers that require technical drawing.
   Answer: Architect, interior designer, engineer, draftsperson, cabinetmaker.

5. What five elements are needed for photography?
   Answer: Light, film, a camera, chemicals, a darkroom.

6. How is a single lens reflex camera different from a viewfinder camera?
   Answer: In a viewfinder camera, you look through a separate viewfinder to compose the picture. In a single lens reflex camera, the single lens is used for both viewing and focusing.

7. How did Johannes Gutenberg improve communication technology?
Answer: He developed a system of mechanical printing using individual letters (movable type), good quality paper and ink, and a screw press.

8. Give an example of something printed by each of the following processes:
   a. relief printing
   b. gravure printing
   c. offset printing
   d. screen printing
   Answer: a. Most typed or word processed letters, business cards. b. Newspaper comics, Sunday newspaper magazine sections. c. Most brochures, textbooks, magazines. d. Decals, posters, T-shirts.

9. How does desktop publishing differ from word processing?
Answer: Desktop publishing composes entire pages consisting of text and graphics; word processing is the combination of the typewriter and the computer, for text only.

10. How is static electricity used in laser printers and copying machines?
Answer: The drum is given a positive charge of static electricity. The toner is given a negative charge and is attracted to the drum. The paper is given a positive charge and pulls the toner to it. Heat and pressure permanently bond the toner to the paper.

CHAPTER 7 REVIEW QUESTIONS AND ANSWERS

1. What is an electronic communication system? Define the term, and give three examples.
Answer: An electronic communication system is one in which the channel carries an electric or electromagnetic signal. Examples include radio, telephone, television, facsimile, microwave links, tape recorders, and compact disks.

2. A newspaper uses electronic typesetting machines to help set its type. Is the newspaper an electronic communication system? Explain your answer.
Answer: The newspaper is not an electronic communication system. In a newspaper, the channel is the ink on the paper that gets delivered to the readers. The electronic typesetter is an aid in setting it up, not the channel to the reader. However, the typesetter can be thought of as a subsystem within the newspaper; this subsystem contains an electronic communication system.

3. Use the Morse code chart provided in the text to write "technology education" in Morse code.
Answer: – · –·–· .... –· ––– ·–·· ––– ––· –·––
              · –·· ··– –·–· ·– – ·· ––– –·

4. Are telephone circuits analog or digital? Explain your answer.
Answer: This question was asked previously in Chapter 4. Based on the knowledge the students had after reading Chapter 4, the most likely answer was "analog". Based on the material presented in Chapter 7, it can be seen that there are advantages in converting the analog telephone signals into digital form for switching and long distance transmission. Using digital telephone circuits, telephone signals are freer of noise and interference.

5. What two products were the result of combining radio and telephone technology?
Answer: Some of these technologies include satellite communications, cellular radios used in car and portable telephones, and cordless telephones.

6. A radio signal has a frequency of 30 MHz (million cycles per second). What is its wavelength in meters?

Answer: Wavelength is equal to the velocity of propagation divided by the frequency of the wave (see page 174). The wavelength is thus:

$$\text{Wavelength} = \frac{\text{Speed of Light (meters per second)}}{\text{Frequency (cycles per second)}}$$

$$\text{Wavelength} = \frac{300,000,000 \text{ meters per second}}{30,000,000 \text{ cycles per second}}$$

$$\text{Wavelength} = 10 \text{ meters}$$

7. Do microwave radios operate at a higher or lower frequency than FM radios?

Answer: Microwave radios operate in the region of one to thirty GHz (one thousand to thirty thousand MHz), while FM broadcast radios operate in the region of one hundred MHz (see spectrum chart on page 177). Microwave radios therefore operate at a higher frequency than FM broadcast radios.

8. Why is fiber-optic cable better than copper wire for large numbers of telephone circuits?

Answer: One of the biggest advantages of fiber-optic cable over conventional copper cable is that many more telephone conversations can be carried on fiber-optic cable, resulting from the larger system bandwidth offered in a fiber system. A second advantage is that fiber-optic cable does not pick up electrical noise (as does copper cable), resulting in crisp, noise-free telephone conversations.

9. How has communication technology changed the way computers are used?

Answer: Fast, efficient data communications allow different size computers in different locations to share large computing tasks, producing more computing power, storage space, and other resources than would be available in a single computer. This spreading out of computing resources is called distributed computing.

10. What is a modem? How does it work?

Answer: A modem is a device that enables data (digital) communications to take place over telephone (analog) lines by converting the data into analog signals (modulation) at the transmitting end, and back to data signals (demodulation) at the receiving end. Modems are used to allow computers to communicate over telephone lines.

11. Why does stereo sound more real and natural than single-speaker sound?

Answer: Because we have two ears, we are used to hearing sounds come from two directions. Stereo sound recreates the two-directional quality of the original sound, while a single-speaker system (monophonic) can only recreate the combined sounds from all directions at one place.

12. Would you like to be a telecommuter? Why or why not?

Answer: Each student should give a personal perspective on the advantages or disadvantages of working at home and communicating with the office with data communication equipment.

CHAPTER 8 REVIEW QUESTIONS AND ANSWERS

1. Why must materials be processed?

Answer: To make them more useful, and to increase their value.

2. Give an example of a food that is processed from animal or vegetable material. Give two reasons foods are processed.

Answer: Examples include cheese, milk, and meat. They must be processed (for example, by cooking) to kill bacteria and make them safe for consumption.

3. What happens to materials in a technological process?
   Answer: The seven resources are combined.

4. What is the role of computers in processing materials?
   Answer: Computers can be used to control machines and flow of materials, and to adjust production processes.

5. What are four ways materials are processed?
   Answer: Forming, separating, combining, and conditioning.

6. What processes would you use to make hamburgers from raw meat?
   Answer: Processes such as: a. forming the hamburger patties; b. separating enough meat for patties from the whole package of meat; c. combining spices with the meat; d. conditioning the meat by cooking it.

7. How do sawing, drilling, and grinding differ?
   Answer: Sawing is cutting material with a blade that has teeth, while drilling is used to cut holes in materials, and grinding is removing small amounts of material at a time using tiny pieces of very hard material called abrasives.

8. What kind of fasteners would you use to attach a metal bracket to a wooden shelf?
   Answer: Wood screws; epoxy resin.

9. Why would you choose to weld two pieces of metal instead of gluing them or using screws?
   Answer: Welding is stronger.

10. Why might you use a different finish on wood than on metal?
    Answer: Wood finish generally allows the natural beauty of the wood grain to be seen, and often enhances it. Metal finishes like paint and enamel obscure the surface. Some metal finishes like clear lacquer are used to protect metal surfaces while still permitting the natural surface to be seen. Such finishes are generally used on jewelry or art objects made of precious metals like silver or gold.

11. What happens to the internal structure of a piece of steel when it is magnetized?
    Answer: The atoms of the steel form magnetic domains and line up with the north poles all facing in the same direction.

12. List five ways to condition materials.
    Answer: Hardening, tempering, annealing, freezing, firing, magnetizing, chemical conditioning, cooking, exposing photosensitive material to light.

13. Make a list of ten jobs in which materials are processed.
    Answer: Product engineering and design, research and development, fabrication, assembly, sales, marketing, management, machine maintenance and repair, machine operation, tool and die construction, and inventory and transportation of goods. Carpenters, masons, plumbers, and electricians.

14. Explain why materials are chosen for products on the basis of their properties.
    Answer: The properties of a material determine its ability to withstand forces or stresses or to conduct heat, light, or electricity. Materials that are needed for a specific purpose must have characteristics that will match their intended use.

15. We plan to manufacture a jigsaw puzzle for four-year-olds. We want to sell it for under two dollars. What material might we make it from?
    Answer: a. Paper (cardboard, masonite); b. plywood; c. plastic; d. rubber; e. clay.

16. You need a screwdriver for electrical work. From what kind of material should the handle be made, and why?
    Answer: Plastic, because it is strong, rigid, and a good insulator. Never metal, because metal conducts electricity.

17. Why must people think about the disposal of end products when choosing the materials from which they are made?
Answer: We live in a "throw away" society; instead of repairing something, we throw it away. When we throw away products, we either bury them in landfills or burn them. But the materials these products are made of can cause pollution. Burning then can pollute the air. Storing them in landfills can pollute groundwater, and besides, landfills are overflowing. Toxic materials — lead and mercury, for example — and radioactive waste must be disposed of in a way that does not pollute the soil, water, or atmosphere. Materials that will decompose should be used or materials should be able to be recycled.

CHAPTER 9 REVIEW QUESTIONS AND ANSWERS

1. Give five examples of how manufacturing technology has helped satisfy people's needs and wants.
Answer: The need for food has been satisfied by manufactured products like processed and preserved foods and new agricultural varieties. The need for clothing has been satisfied by items manufactured from natural fibers, and synthetics. The need for medical care has been satisfied by the manufacturing of pharmaceuticals and instrumentation. The need for transport has been satisfied by the production of air, land, and seagoing vehicles. The desire for recreation and exercise has been satisfied by the production of sporting goods equipment.

2. Explain how craft production differs from mass production.
Answer: See chart on page 253.

3. How did mass production improve people's standard of living?
Answer: Mass production and the standardization of parts brought prices down so that the masses were able to afford items that were before only within reach of the wealthy.

4. Draw a flowchart of an assembly line for making a greeting card.
Answer: See following illustration.

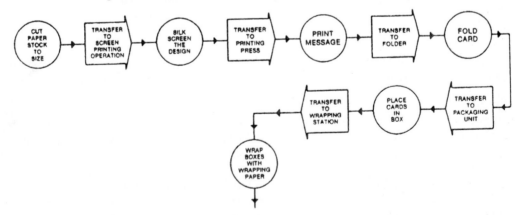

5. What are two disadvantages of mass production?
Answer: Workers do not get the satisfaction of producing a complete product; Factories are expensive to equip; Product modifications may require major retooling costs; Work is often routine and dull.

6. What are the two subsystems that make up the manufacturing system?
Answer: The material processing system, and the business and marketing system.

7. Suggest an invention that would help you with your homework.
Answer: Leave this to the students' imagination.

8. How might you innovate a soda can?
   Answer: Leave this to the students' imagination.

9. What would be a suitable product for your technology class to manufacture?
   Answer: This depends upon the types of facilities available. Keep costs down. It will be easier for the class to sell inexpensive items for a profit. Typical items might be: Wooden Toys; Key Chains with School Logo; Agricultural Products; Hand Creams and Lotions; Printed Greeting Cards; Food Items; Items of Clothing, like T-Shirts; and Headbands with Logos.

10. If you could form a quality circle with your classmates and teacher, what improvements would you suggest be made in the technology laboratory?
    Answer: Improvements could focus on safety, rearrangement of machinery for easier work flow, aesthetic redecoration, new and better displays, and repair of work surfaces and/or tools and machines.

11. Why do manufacturers want their products to be uniform?
    Answer: So that parts are interchangeable; so that high quality will be maintained; so that returns and repairs will be minimal; so that the company maintains a good reputation with its customer base.

12. What could be two undesirable outcomes from a system that manufactures computers?
    Answer: More people might be put out of work than the number of new jobs created; Computers could become so pervasive that work involves less human contact and little socialization.

13. How is feedback used to ensure good quality in manufactured products?
    Answer: The output of the process is monitored, and compared to a desired standard. If discrepancies are found, changes are made in the process.

14. Do you think that robots should be used instead of people on assembly lines? Explain why or why not.
    Answer: This is a subjective question and asks for student opinion.

15. How have computers affected the manufacturing industry?
    Answer: Computers are now used to control machines that people controlled in the past. Computers also are used to model product designs, inventory materials, keep accounts, and provide automated office functions.

16. Draw a system diagram for a system that manufactures chewing gum. Label the input command, resources, process, output, monitor, and comparison.
    Answer: See following diagram.

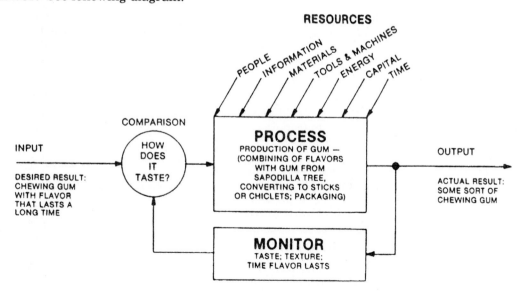

System diagram for manufacturing chewing gum

CHAPTER 10 REVIEW QUESTIONS AND ANSWERS

1. What is the major difference between manufacturing and construction systems?
   Answer: Manufacturing refers to producing products in a workshop or factory; construction refers to producing structures on a site.

2. Draw a labeled systems diagram of the construction system.
   Answer: See diagram, page 279.

3. Describe four career opportunities provided by the construction industry.
   Answer: Architects design buildings and develop the overall site and building plans. Civil engineers prepare exact drawings and plans for the building framework and foundation; they are responsible for the technical parts of the project. General contractors own their own construction companies; they hire workers and are responsible for building part or all of the construction project. Estimators prepare cost proposals called bids, which are the estimates of the cost of the construction work. Tradespeople work on the projects. Some tradespeople work as carpenters, electricians, plumbers, or cement masons. Some operate heavy equipment like bulldozers.

4. How could you finance the construction of a private home?
   Answer: Generally, private homes are financed through a mortgage. A mortgage is a loan that is received from a bank or finance company. The mortgage is paid back over a period of years with interest added to each payment.

5. If you were to choose a site for a movie theater, what are five things you would have to consider?
   Answer: 1. Location (is site convenient to residential areas); 2. Competition (how many other theaters are nearby); 3. Cost of land; 4. Yearly taxes; 5. Conditions of the ground; 6. Ease of access to site; 7. Ease of waste disposal.

6. What kind of superstructures do the following structures have?
   a. the Washington Monument
   b. a tower that supports electrical wires
   c. a skyscraper
   d. a large dam
   e. a castle from the Middle Ages
   Answer: a. the Washington Monument — mass superstructure; b. a tower that supports electrical wires — framed superstructure; c. a skyscraper — framed superstructure; d. a large dam — bearing wall superstructure; e. a castle from the Middle Ages — mass superstructure

7. Make a sketch of a suspension bridge.
   Answer: Students may use the photograph on page 287 as a model, find other examples in magazines, or they may wish to visit a suspension bridge and sketch it on site.

8. Explain why an arch can support a great amount of weight.
   Answer: The arch transfers weight to the supports at its base, which rest upon the ground.

9. Design a tower, using rolls of newspaper as your building material.
   Answer: Permit students to use their own creativity.

10. Name five different types of structures.
    Answer: Bridges, buildings, tunnels, roads, airports, canals, dams, harbors, pipelines, and towers.

CHAPTER 11 REVIEW QUESTIONS AND ANSWERS

1. List three factors to consider when choosing a lot on which to build a house.
   Answer: A builder considers the ease with which utilities and other services (like roads) can be brought to the location. Cost of land is important and will be influenced by such things as the view and the proximity to metropolitan areas. How the house will be placed on the lot is also important (energy conservation).

2. How are the footing and the foundation wall alike? How are they different?
   Answer: Both the footing and the foundation wall are part of every house foundation. The footing is the base of the foundation and spreads the weight of the structure over a wider area of ground, while the foundation wall supports the whole weight of the structure and transmits it to the footing.

3. Why are slab foundations well suited to warmer climates?
   Answer: Slab foundations are suited to warmer climates where the frost line is not far below grade.

4. What are two purposes served by the walls in a house?
   Answer: Walls transmit the weight of the structure to the foundation (load bearing walls). They also serve as partitions between rooms (partition walls).

5. Sketch a section of framed wall that includes one window opening. Include and label the studs, top and bottom plates, headers, trimmers, and jack studs.
   Answer: See drawing on page 311 (framing a wall section).

6. Why is sheathing nailed over the entire wall, including windows and doors, when it is installed?
   Answer: It is more economical in the long run because it saves on labor costs.

7. Why is a trussed roof faster to build than a frame roof?
   Answer: A trussed roof is one that is mostly prefabricated and trucked to the construction site. A frame roof is built on site.

8. Think about a house you might design for yourself. What kind of roofing material would you use?
   Answer: Asphalt shingles (they are weatherproof and reflect the sun's heat); wooden shingles or shakes (they are very attractive).

9. How does insulation increase comfort and cut down on cost?
   Answer: Insulation keeps the inside of the house isolated from the outside. It prevents heat from escaping during cold winter days, and prevents heat from entering the house during hot summer days, thus keeping the temperature of the house constant.

10. What purpose does the vapor barrier on the outside of insulation serve?
    Answer: It prevents warm air from reaching the colder side of the house and protects against condensation.

11. Why are there separate systems for fresh water and waste water?
    Answer: So that the waste water does not contaminate the fresh water supply.

12. Explain how a home heating system works. Use a simple system diagram as part of your explanation.
    Answer: See diagram below. You may also want to have students review the systems concept in Chapter 3.

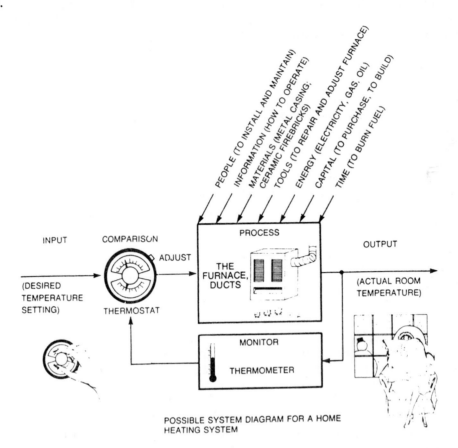

POSSIBLE SYSTEM DIAGRAM FOR A HOME HEATING SYSTEM

13. Are circuit breakers connected in series or in parallel with the electrical outlets in a room? Explain your answer.
    Answer: The circuit breaker must be connected in a series with the circuit it is protecting. Since a series circuit has only one path, if the circuit breaker trips, the entire path is broken.

14. How does grounding electrical equipment guard against electrical shock?
    Answer: In the case where a wire carrying electricity touches the metal casing of a piece of equipment, a ground wire provides a better path for electricity than the person holding the equipment. The electricity will take the path of least resistance. Instead of flowing through the person, it will flow through the ground wire.

15. What are some of the differences between commercial buildings and houses?
    Answer: Commercial structures serve more people (and thus demand more capability from utility services). They are generally taller, and must be designed to withstand higher wind loads.

16. Why does wind affect tall buildings to a greater degree than it affects short buildings?
    Answer: The wind load on a building increases as the square of the building height.

CHAPTER 12 REVIEW QUESTIONS AND ANSWERS

1. What part does management play in a production system?
    Answer: Management's part in a production system is to coordinate the proper use of resources within the company. Management also decides what products to make or what projects to build, and how to make potential customers aware of the company's products.

2. What do architects do on a construction project? What do engineers do?
Answer: Architects design a building's shape and select basic material for it. Engineers design the building's structure and its major systems.

3. List at least three tasks assigned to the general contractor.
Answer: A general contractor is a person or company who takes the overall responsibility for a construction company. The contractor must do the many jobs of management. People must be hired to work on the building. Work must be scheduled so the many jobs are done in the right order. Materials must be purchased. Delivery must be scheduled near the time they are to be used. These are all tasks assigned to the general contractor.

4. What are subcontractors?
Answer: Subcontractors are hired by the general contractor to perform specific job tasks.

5. Why must a contract between the general contractor and the owner be so long?
Answer: The contract includes the architect's and engineer's drawings. It tells in detail what will be done, when it will be done, and who will do it. It also usually tells who is responsible if there is an accident, and who must get the building permits. There may also be penalties that will come out of the contractor's pay if the job is not finished on time.

6. How is the job of a project manager different from that of a general contractor?
Answer: On very large construction projects, a project manager acts as an independent coordinator for all job tasks. The project manager works for the owner, and oversees the many subcontractors, as well as the general contractor.

7. What is coordination between subcontractors on a construction job? Why is it important to getting the best quality work done, and the job finished on schedule?
Answer: Coordination is the process of making sure that the work of one subcontractor does not interfere with the work of another subcontractor. Specific jobs must be done in the right order and done properly before other work can be completed. If the job is not finished on schedule, the price of the project may be driven up. These additional costs are often called cost overruns.

8. How does a Gantt chart help in keeping a construction project on schedule?
Answer: A Gantt chart is used to help mangers schedule and monitor a project. The chart shows when work is to start and when it is to be finished.

9. What are the four basic tasks that all managers do?
Answer: Planning, organizing, leading, and controlling.

10. What is the difference between marketing and sales?
Answer: Marketing is the task of deciding what potential customer groups the company should develop products for, and what products to develop. Sales is the actual task of selling the company's products to its identified customer groups.

11. How are research and development different?
Answer: Research is the search for new materials, processes, and techniques that may be useful to a company's business. Development is the use of these new discoveries and previously known methods and techniques to solve a specific problem.

12. Is quality control in the manufacturing stage important to the jobs of marketing and sales? Explain your answer.
Answer: The quality of a product is one thing that can set it apart from other similar products. Customers will compare the quality of products from different manufacturers, along with functionality and price, when deciding which product to buy. Quality control is thus important to both sales and marketing.

13. Name four ways a manufacturer might sell products.
    Answer: A manufacturer can sell products through direct sales, through sales representatives, through dealers, and through distributors.

14. What three things should be included in a business plan?
    Answer: The business plan should include; the goals and objectives of the business, the strategy and methods that will be used to achieve these goals, and the financial requirements for the business.

15. Draw a PERT diagram for making hot cocoa, from the purchase of ingredients through cleaning up afterward.
    Answer: The students should draw a PERT chart from information provided by the instructor or from information that the students generate with the guidance of the instructor. Students should refer to the chart on page 356.

16. If it takes 90 days for a company to produce and sell a product, what is the approximate turn on the product?
    Answer: Ninety days is about three months. The manufacturer can produce and sell four batches of products in one year. The turn on the product is therefore about 4.

CHAPTER 13 REVIEW QUESTIONS AND ANSWERS

1. A girl pushes on a bicycle with 10 pounds of force, moving it 100 feet. How much work has she done?
   Answer: She has done 10 pounds x 100 feet = 1000 foot-pounds of work.

2. A boy pushes on a car with 25 pounds of force, but does not move it. How much work has he done?
   Answer: He has done no work, because the car has not moved any distance.

3. While playing pinball, a boy draws back the spring-loaded shooter to put a ball into play. At what point does the shooter have potential energy? At what point does it have kinetic energy? At what point does the ball have kinetic energy?
   Answer: When he draws back the shooter, kinetic energy produced by the muscles in his arm stores potential energy in the shooter. When he lets the shooter go, the potential energy is converted to kinetic energy. The ball has kinetic energy when it is moved by the shooter into play.

4. Will the energy generated by falling water in a hydroelectric plant be more, less, or the same as the gravitational energy stored in the water that falls (assuming there are no losses in the generator)? Explain your answer.
   Answer: The same: Energy cannot be created or destroyed, but it can be changed from one form to another.

5. Is solar energy really unlimited? Why do we call it an unlimited energy resource?
   Answer: Since the sun is expected to burn itself out in about two billion years, the sun is not truly an unlimited source of energy. We call it unlimited because for all practical purposes, during our lifetimes and those of many generations to come, it seems unlimited.

6. How is energy stored in a fossil fuel?
   Answer: Energy is stored in the high-energy bonds of the molecules. When the fuel is burned, the large molecules are broken down into smaller ones, releasing some of the energy in the bonds.

7. Name three fossil fuels.
   Answer: Three major fossil fuels are oil, natural gas, and coal.

8. Why are researchers trying so hard to solve the problems of using coal for energy?
Answer: Coal is the limited energy source of which we have the most, but there are problems with its use. Coal must be moved in large quantities by train, barge, or truck to its destination. Burning coal also produces sulfur dioxide, which is an air pollutant.

9. How is nuclear fission different from nuclear fusion? Why is fission considered a limited energy source while fusion is considered an unlimited energy source?
Answer: a. In nuclear fission, energy is released when a large atomic nucleus is bombarded by neutrons and split into two smaller nuclei, more neutrons, and energy. In nuclear fusion, energy is released when two small nuclei are forced together to form one larger nucleus.

b. Fission is classified as a limited energy source because the large nuclei needed for the process are found in uranium, a fuel that there is a large but finite supply of. Fusion is considered an unlimited energy source because the small nuclei needed for the process are found in sea water.

10. Why would a solar-powered car be an impractical means of transportation?
Answer: This might be an impractical means of transportation because it would not work in the dark (at night, or in tunnels or under bridges), and because a very large number of solar cells would be needed to move realistic loads.

11. Is wood used for home heating in your area? Why or why not?
Answer: The answer will depend on your area. Reasons for not using wood will include transportation, storage, local availability, etc..

CHAPTER 14 REVIEW QUESTIONS AND ANSWERS

1. What is power? How is it different from work?
Answer: Power is the time rate of doing work, or the amount of work done during a given period of time. The concept of work involves only force and distance; it does not include any consideration of amount of time needed to move an object.

2. Name the two major parts of a power system. Describe what each part does.
Answer: The two major parts of a power system are the engine and the transmission. The engine uses energy to create mechanical force and/or motion. The transmission transfers force or movement from one place to another, or changes its direction.

3. List four different kinds of engines.
Answer: External combustion engines, internal combustion engines, reaction engines, and nuclear reactors. Electric motors are also engines when used in this context.

4. Describe the difference between an internal combustion engine and an external combustion engine. Give one example of each.
Answer: In the internal combustion engine, the fuel used is also the expanding gas; in the external combustion engine, the fuel is separate from the gas that expands. Internal combustion engines are used in cars, trucks, some airplanes, many railroad locomotives, and ships. Small engines are used in lawn motors and go-carts. Also portable generators, leaf blowers, snow throwers, and outboard engines. External combustion engines are steam engines.

5. How is a gasoline engine different from a diesel engine?
Answer: A diesel engine does not have a spark plug in the cylinder. Not described in great detail in this text, a diesel engine cylinder also has a fuel injector rather than an intake valve.

6. Describe the steps by which a four-stroke engine turns the energy in fuel into the mechanical energy that moves an automobile.
Answer: The operation of a four-stroke engine is described on page 406.

7. State Newton's third law of motion. How is the third law related to reaction engines such as rocket and jet engines?
Answer: Newton's third law states that for every action, there is an equal and opposite reaction. In both rockets and jets, the force of the hot gas rushing out the exhaust (action) pushes the engine, and therefore the vehicle, forward (the reaction).

8. What is an electric motor? List four devices that contain an electric motor.
Answer: Electric motors are electromagnetic devices. They change electrical energy into rotary motion. Small motors can be found in appliances, toys, and small machines. Large motors are used to move subways, trains and elevators.

9. Tell how a nuclear power plant produces electricity.
Answer: In a nuclear power plant, the heat generated by atomic fission (the splitting of heavy atomic nuclei into smaller nuclei, neutrons, and energy in the form of heat) is used to boil water, making steam. The steam turns a turbine, which in turn turns a generator. The generator converts the rotary motion into electricity. The steam is then cooled back to water and starts the cycle again.

10. Why is it necessary to send electric power at high voltage when it is sent over long distances?
Answer: When power is transmitted at high voltage, voltage losses (often called "line drops") are lower, and an even smaller percentage of the total, than when power is transmitted at low voltage. Overall losses are thus lower, and efficiency higher, when sending electric power at high voltage.

11. Why would a superconductor that works at high temperatures be useful in an electric power transmission system?
Answer: Transmission lines made of superconducting materials will have little or no loss of electricity as it flows through the wires. Lines that do not lose power will mean less costly electricity.

12. Why is water sometimes pumped back up to the other side of a dam in a hydroelectric plant?
Answer: Water is pumped back up over the dam during periods of low demand by customers. The water is then available again during the next period of peak demand, when it will be used to fall through the turbine again.

13. What is a battery? How does it work?
Answer: A battery is a device that stores chemical energy and converts it to electricity (DC). Current is made to flow by the chemical reaction between electrodes and electrolyte.

CHAPTER 15 REVIEW QUESTIONS AND ANSWERS

1. Describe how transportation and communication systems have made countries interdependent.
Answer: Communication systems bring about an exchange of ideas between people of different countries; transportation systems allow countries to buy and sell goods from each other, and permit people from one country to travel in and meet people from other countries.

2. In what way(s) have modern transportation systems affected the way your family lives?
Answer: Answers will, of course, be individualized, but should center around the ease of going places by car, railroad, mass transit, or airplane.

3. Is an automobile a necessity or a luxury in your family? Why?
Answer: Again, individual answers will vary, but the general pattern may be that in an urban environment, cars may be luxuries or unnecessary, while in suburban or rural areas, cars may be necessities.

4. Why were steam engines replaced by internal combustion engines in cars?
Answer: Internal combustion engines can go further without needing more fuel and/or water; the steam under pressure is also a potential problem in an accident (so, also, is spilled gasoline).

5. Why were steam engines replaced by diesel and electric engines in trains?
   Answer: Diesel and electric engines run more efficiently, cause less air pollution, and are easier to maintain.

6. What role did trains play in the settling of the American West?
   Answer: The completion of the transcontinental railroad in 1869 made it easier to move people and goods to and from the West, speeding the pace of settlement.

7. Why are electric engines used on railroads that travel into major cities?
   Answer: Electric engines are used because they do not pollute the air where they are operating. (It should be noted that the generating plant that produces the electricity in the first place may pollute the air in a different location.)

8. Describe why a boat made out of steel and cement can float.
   Answer: The boat's hull is made of such a shape that air fills the hull. As long as the combined weight of the hull (steel and cement) and the air contained in it is less than the weight of the water displaced (pushed aside) by the hull, the boat will float (See page 434).

9. Describe how intermodal transportation works and what its advantages are.
   Answer: In an intermodal transportation system, freight is loaded into sealed containers that are carried by truck, railroad, ship, or airplane from the sender to the final destination. In most cases, some combination of different types of transportation is used. Because the cargo is not constantly loaded and unloaded, but is always sealed in a container, it is less likely to be damaged or lost. Because different modes of transportation are used, the most efficient mode is used for each portion of the trip.

10. What shapes do submarines and submersibles use? Why?
    Answer: Ships that travel underwater are circular or cylindrical in shape to provide the greatest possible strength to withstand the high pressures encountered underwater.

11. Describe how a wing enables a plane to fly.
    Answer: Air traveling over the curved top surface of the wing area creates an area of low pressure, while air traveling along the straight bottom surface of the wing creates a higher pressure; the two combine to put an upward force (lift) on the wing.

12. Give at least two reasons why jet engines have replaced internal combustion engines on commercial passenger planes.
    Answer: Jet engines have fewer moving parts and require much less maintenance than internal combustion engines; they also move planes at faster speeds, lowering travel time.

13. How is a rocket different from a jet engine?
    Answer: While the two work on the same action-reaction principle, a jet engine uses the surrounding air to burn fuel, while a rocket carries its own oxidizer (often liquid oxygen).

14. Why are rocket engines used for space vehicles?
    Answer: Rockets are used for space vehicles because they produce enough thrust to propel the vehicle into orbit (or beyond), and because they carry their own oxygen (in space, there is no oxygen to burn fuel with).

15. Name two kinds of transportation systems that don't use vehicles. Describe how they work.
    Answer: Two kinds of nonvehicle transportation systems are pipelines and conveyors. In a pipeline, pumps move fluid along the line, sometimes over distances of thousands of miles. Conveyors carry objects from place to place by having them ride on a belt or series of closely-spaced wheels.

16. Name a human-powered vehicle that is widely used for sport or leisure travel in this country, but that is used as basic transportation in other countries.
    Answer: The bicycle.

CHAPTER 16 REVIEW QUESTIONS AND ANSWERS

1. What four possible combinations of outputs can result from technological systems?
   Answer: Expected and desirable, expected and undesirable, unexpected and desirable, and unexpected and undesirable.

2. Do you believe that technology is good, evil, or neutral? Explain why.
   Answer: Technology, in and of itself, is neither positive nor negative; it is the way people use technology that determines whether it is beneficial or harmful to humans and the environment.

3. How can people control the development of a technology they feel may be harmful?
   Answer: By modifying the technological process so that it achieves the positive purpose without the negative impacts. By projecting possible negative outcomes and developing technological solutions to reduce them; by legislation that restricts or prohibits development; by voting to restrict or prohibit implementation.

4. Give two examples showing that pollution is not just a recent problem.
   Answer: Rivers around ancient Rome were so polluted that people were forbidden to bathe in them; a century ago, horses in New York City produced over 500,000 tons of manure each year; garbage piled up in the streets; industries billowed clouds of black smoke into the atmosphere.

5. Give an example of technology that is poorly matched to the human user.
   Answer: The typewriter keyboard (the most commonly used letters are not within easiest reach of the fingers); light bulbs that glare; stereo headsets that are uncomfortable; auto dashboards that do not provide a clear view of the instruments; pesticides that cause illness; shampoos that burn eyes; traffic lights where the yellow light stays on for too long or too short a time.

6. Give an example of a problem technology has caused and a solution technology has provided.

Answer: PROBLEM	SOLUTION
automobile pollution	the catalytic converter
high energy consumption	alternative energy sources
crowded living conditions	skyscrapers
high demand for products	automated mass production
production of carcinogens	anti-cancer drugs
crowded roadways	superhighways

7. Explain how technology can make life easier for a disabled person.
   Answer: Hearing aids, eyeglasses, and contact lenses can help those who are hearing or visually impaired. Biotelemetry helps analyze abnormal muscle activity. Adaptive devices can be made that fit the needs of people with physical disabilities. Buses are made to kneel at curbside; devices are available that transport people in wheelchairs up stairs; automobiles are modified so that they can be driven by people who do not have the use of their legs; see page 468 (Handbikes and Sunbursts).

8. Draw a design for a futuristic communication system that would allow you to communicate with a class of students in Europe.
   Answer: Leave this to the imagination of the students.

9. What would be two good products to manufacture in space? Why?
   Answer: Drugs, because they could be made to be very pure (homogeneous crystalline structure); semiconductor crystals; glass (see pages 474–476).

10. How will future travel differ from present-day travel?
    Answer: Possibilities include automated computerized control systems for cars; increased use of people-movers (moving sidewalks, PRTs); transatmospheric (hypersonic) aircraft; personal space vehicles.

11. Would a personal rapid transit (PRT) system be useful in your town or city? Explain why or why not.
Answer: Over short distances in areas with high population density, PRTs would be useful. In rural areas, conventional means of transit would be more cost-effective.

12. What kinds of moral questions will arise if people are able to engineer human life?
Answer: Should people have the right to alter the makeup of the human life form? In whose hands would the power lie? Could governments use this knowledge to engineer a superhuman race of fighters to take over other nations? Who will decide who is selected for improved qualities and an extended life? If humans are engineered to live longer lives, will we face crushing overpopulation?

13. Give an example of how a modern or future technology will require inputs from several existing technological systems. (**Hint:** Think of how biotechnical, communication, construction, manufacturing, and transportation systems act together to produce new technologies, such as manufacturing in space.)
Answer: Guided missiles and satellites use communications, construction, manufacturing, and transportation technologies; genetic engineering uses manufacturing, information processing, and biological technologies; colonies in space will need biological, communication, construction, manufacturing, and transportation technologies.

14, Draw a futures wheel. At the center, place a skin cream that brings back youth. Predict the possible outcomes.
Suggestion: Use the futures wheel on page 484 as a model, but leave this to student's imagination. Possible outcomes should reflect some undesirable outcomes, such as overpopulation, too many workers for the amount of jobs, etc..

APPENDIX C CROSSTECH PUZZLES

PREPARATIONS

Preparations for use of the CROSSTECH puzzles for *LIVING WITH TECHNOLOGY Teacher Resource Book* include the following:

1. Use the master to make copies of the CROSSTECH puzzle for class members.
2. Prepare Overhead Transparencies of both the CROSSTECH puzzle and the answer sheet.
3. Modify the CROSSTECH puzzle for students requiring learning modifications.

USES

Suggestions for use of these puzzles include:

1. Spelling and vocabulary study
2. Pretest of chapter to assess student knowledge of chapter content
3. Review of chapter material
4. Test in lieu of the chapter test
5. Retest of chapter material

METHODOLOGY

Learning styles and abilities can vary within and among the classes. Time constraints can dictate the way in which a learning instrument such as the CROSSTECH puzzle is administered. Here are a few methods that can be incorporated in the lab.

1. Homework assignment
2. Individually assigned as part of classwork
3. Cooperative assignment: Partners divide up the puzzle and complete one part and teach this part to his or her partner. The partner in turn completes and teaches his or her part of the puzzle.
4. Class drill: use words and quiz class as a review of information to be covered in a chapter test.
5. Spelling test: use these words to introduce students to technological terms and their correct spelling.
6. Class discussion: Students volunteer and explain answers as the entire class completes the CROSSTECH puzzle together.
7. Computer drill: Using crossword puzzle software, enter the crossword puzzles into the computer and have students complete the puzzle on the computer.

CHAPTER 1
TECHNOLOGY IN A CHANGING WORLD

ACROSS

1. The study of why natural things happen the way they do.
5. Technology is a process that uses many kinds of _____ to meet people's needs.
7. Henry Ford is called the "father of _____ production."
10. The use of satellites is a form of _____ technology.
11. Modern technology produces _____ using oil, gas, coal, and nuclear fuels.
12. The use of knowledge to turn resources into goods and services that society needs.
15. The period of time or age when inventions are based on electronics and the computer.

DOWN

2. The era when mechanical devices and machines were invented.
3. A period of time when people lived off the land. Discoveries and tools related to harvesting.
4. Our _____ and the way we live are affected by the products of technology.
6. The rate of change of technology at the present time is known as _____.
8. The production of cars in an automobile plant is known as _____.
9. Stones, bones, and wood were used for tools during the _____ Age.
13. People made tools and weapons from a mixture of copper and tin during the _____ Age.
14. The age beginning around 1200 B.C. when iron came into common use.

CHAPTER 2
RESOURCES FOR TECHNOLOGY

ACROSS

3. Any form of wealth such as cash, stock, buildings, machinery and land.
6. Substitutes for some resources are known as _____ materials.
10. Heat from deep inside the earth provides _____ energy.
11. Conversion of radioactive matter into energy is called _____ energy.
12. A type of a simple machine which has a surface placed at an angle to a flat surface.
13. A resource that can range from less than one billionth of a second to human lifetimes.
14. A form of energy which provides wind, heat, and light energy.
16. A resource of technology creating and using the products and services of technology.

DOWN

1. A nonrenewable raw material which is a source of energy.
2. Resources that change the amount, speed, or direction of a force.
4. An optical tool that sends bursts of light energy that can measure, cut and weld materials.
5. People, information, materials, machines, energy, capital and time used in technological systems.
7. A resource found in computer files, books, films, and museums.
8. The resource which includes nuclear, solar, gravitational, tidal, and geothermal.
9. A natural resource found in nature such as plants or minerals is a raw _____.
15. A resource classified as hand or machine which extends human capabilities.

CHAPTER 3
PROBLEM SOLVING AND SYSTEMS

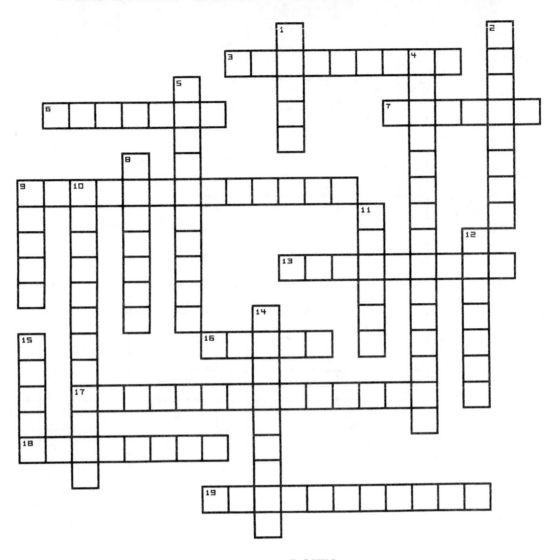

ACROSS

3. Model or final product built to help evaluate a design of a product and check for problems.
6. Systems with feedback are sometimes called _____ systems.
7. A design _____ is a record of information, drawings, and ideas used in solving a problem.
9. A creative way of thinking in which individuals in a group suggest ideas.
13. Systems that can be combined to produce powerful systems.
16. The command we give a system or the desired result.
17. Trying out the proposed solution in problem solving.
18. The use of information about the output of a system that can be used to modify the process.
19. A type of system that has feedback.

DOWN

1. A full sized or smaller scale version of a proposed solution to a problem.
2. Market _____ is done to determine if customers like a new product.
4. Multistep process used to reach solution in response to a human need or want.
5. The process of observing the output of a system.
8. Coming up with alternative solutions by thoroughly researching the problem.
9. A type of research that deals with the nature of different materials and processes.
10. Different ideas for solutions to problems.
11. Means of getting a desired result through processing resources in response to a command input.
12. _____ results are a system's command input; a statement of the expected output of a system.
14. A system that does not use information about the output to affect the process.
15. A design _____ states in very general terms the solution one thinks might solve the problem.

CHAPTER 4
THE ELECTRONIC COMPUTER AGE

ACROSS

1. Controls the computer memory, output devices, and use of software, DOS is an example.
3. Binary digit (abbreviation).
6. A material (like silicon) that is neither a good conductor nor a good insulator.
9. Information that is coded into a series of 0s and 1s.
10. The force that makes current flow through a conductor.
11. An electronic component that allows a small amount of current to control a larger current.
14. The heart of the computer system. It controls the flow, storage, and manipulation of data.
15. A material that permits electrons to flow through it.
16. A material that does not conduct electricity.

DOWN

2. A complete circuit on a semiconductor chip.
3. Eight bits.
4. A signal that is similar to the information it represents.
5. The fastest and largest kind of computer.
7. A flow of electrons through a conductor.
8. Opposition to electron flow, measured in ohms.
12. A set of instructions that tells a computer to perform a task.
13. A combination of electronic components.

CHAPTER 5
COMMUNICATION SYSTEMS

ACROSS

1. A category of communication systems where the channel carries electronic signals.
4. A form of communication where a computer is connected to a machine.
6. A goal of a communication system used to convince people to do something.
7. Interference which affects our ability to clearly understand the message.
8. The part of a communication process that actually sends the message.
11. Includes having a message sent, received, and understood.
13. A goal of communication with the purpose of amusing the audience.

DOWN

2. The route a message takes.
3. The part of a communication system that accepts the message.
5. Forms of communication such as radio, television, or newspapers, used to reach the general public.
9. A goal of communication with the purpose of teaching people.
10. A goal of communication with the purpose of giving out facts to people.
12. A type of communication where the channel carries pictures or printed words.

CHAPTER 6
GRAPHIC COMMUNICATION

ACROSS

5. Intaglio or a type of printing used to print some magazines.
6. Drawings done in three dimensions such as oblique, isometric, and perspective.
7. Printing from a flat surface where the image is transferred from a blanket to paper.
10. Computer-aided design or computer-aided drafting.
11. Machine that uses a photographic process to convert computer output to printed text.
13. A pictorial drawing done within a framework of three lines, two of which are drawn at a 30° angle.
14. A type of printing, used to print posters and T-shirts, where ink passes through a screen.
16. A drawing where the front, top, and side are drawn as separate pictures.

DOWN

1. Known as letterpress printing, it is used to print newspapers and greeting cards.
2. A type of printer with "petals" that carry letters, numbers, and printing characters at the tips.
3. This type of drawing uses tools and computers to communicate the true size and shape of objects.
4. This is the most real looking type of technical drawing.
8. Printer that has a rectangular shaped group of pins which can produce letters and drawings.
9. A type of publishing that lets a person turn out a book or newsletter page by page.
12. When film is exposed to light and then developed, a _____ is made.
13. The type of printer that turns out high quality text and pictures quickly.

CHAPTER 7
ELECTRONIC COMMUNICATION

ACROSS

5. First electronic communication system which involved the use of sounds as a code for letters.
6. _____ automation is used to describe the use of computers and communications in the office.
10. A conference connecting people at different locations by the use of telephones.
12. A device used to send data from a computer through telephone lines to another computer.
16. Small data networks within an office that share data and a printer or modem.
17. The transmission of data from a station on earth to a satellite.
18. Communication between computers or between a computer and another device.
19. A type of computing system where small computers can share work with each other or with a larger computer.
20. A number of computers or computer devices joined together.

DOWN

1. Having two sound channels.
2. Two reels of tape enclosed in a plastic case; can be audio or video.
3. The measurement of cycles per second (one cycle per second is called one Hertz.).
4. A signal changed to a different frequency and transmitted to earth from a satellite.
7. Type of cable made of thin strands of coated glass fiber capable of guiding light around corners.
8. To take work home from the office and back to the office by disk or to transfer it digitally.
9. An _____ wave is the flow of energy through space at the speed of light.
11. The areas where sound is stored on a tape.
13. The radio transmission of a signal out to many listeners.
14. The distance covered by one cycle of an electromagnetic or other wave.
15. A device that makes a small electrical voltage larger.

CHAPTER 8
PROCESSING MATERIALS

ACROSS

1. The property similar to elasticity which allows materials to be bent and stay bent.
5. The mechanical property of a material that allows it to absorb energy without breaking.
7. Materials made by combining several materials, such as plywood.
9. A force that pulls on a piece of material.
11. A cutting process using tools and machines that shear, saw, drill, grind, shape, and turn materials.
13. Materials that can be bent without breaking are known as _____ materials.
15. The force that is opposite of tension which pushes on or squeezes a material.
17. Metals that are made up of more than 50 percent iron.
19. Plastics like bakelite and Formica that do not soften when heated.
20. A process such as tempering or hardening that changes the internal property of a material.

DOWN

2. Plastics that soften when heated and can be melted and shaped such as nylon, plastic bags, and PVC.
3. The force of twisting a material.
4. The process of changing a material's shape without cutting, such as casting or forging.
6. The process of applying a finish to a material, such as painting, electroplating, or anodizing.
8. Material that offers very little electrical resistance such as silver or copper.
10. A forming process where softened material is squeezed through an opening, taking on its shape.
12. A separating process where the sharp edge of a blade compresses the material causing it to break.
14. To reuse a material instead of discarding it.
16. A material that can bend and then come back to its original shape and size.
18. A separating process where a blade that has teeth chips away bits of material as it cuts.

CHAPTER 9
MANUFACTURING

ACROSS

3. Until the middle 1900s, many people were workers on these.
5. Parts that are alike are _____.
7. A model of a product that will later be produced in quantity.
8. A place where products are produced in quantity.
9. A computerized system that links design and manufacturing (abbreviation).
12. A system of producing products in quantity using assembly-line techniques.
13. A reprogrammable device that can replace a human in a factory.
14. A type of manufacturing where machine tools can be reprogrammed to do different jobs.
16. A system of producing products in a factory or workshop.
17. This enables us to automatically control a machine tool.
18. A person who comes up with a brand-new idea for a product.

DOWN

1. A system for ensuring that the quality of a product is what is desired.
2. Computer-integrated manufacturing (abbreviation).
4. Manufacturing where parts arrive just when they are needed.
6. A system of using feedback to control the operation of machines.
10. Products that are made to meet an individual's specific needs are _____.
11. A system that includes manufacturing and construction.
15. An organized group of workers.

CHAPTER 10
CONSTRUCTION

ACROSS

3. Details given to a builder.
5. The usable part of a structure.
7. A strong curved support.
8. These are often made of wood and are used to hold concrete while it is hardening.
10. This is made from cement, sand, water, and stones.
12. A very strong material used as a building framework.
13. A structure rests on this.
14. A loan for a construction project.
15. To produce a part of a structure in a factory.

DOWN

1. A material used in road construction.
2. A person who takes charge of actual hands-on construction.
4. A person who draws designs for buildings.
5. A person who lays out a site.
6. A system of building a structure on a site.
9. An ingredient in concrete, made from limestone and clay.
11. A technical expert who is responsible for exact structural plans.

CHAPTER 11
BUILDING A STRUCTURE

ACROSS

3. The base of the foundation which spreads the weight of a structure over a wide area of ground.
5. This permit allows the builder to begin work.
7. This controls the amount of current that can safely pass through household wiring.
8. A type of housing where parts are built in a factory and taken to a construction site and assembled.
11. A material placed within the walls of a structure that does not conduct heat.
15. The vertical parts of a foundation must be this way so that the house on it is not built crooked.
16. The waterproof material used over insulation acts as a _____ to prevent condensation.
17. The depth to which the ground freezes in winter.
18. The horizontal pieces of lumber made from a double layer of 2 x 4's.
19. Cutting pieces of wood to size and fastening them together to form a framework of a building.

DOWN

1. The _____ plate is a horizontal single layer of 2 x 4's at the bottom of the wall.
2. The slope of a roof.
3. The floor frame is built from long boards spaced about 16 inches on center called _____.
4. Built on top of the footing, it supports the entire weight of the structure.
6. Made of wood or steel and placed down the middle of the floor supporting joist sections.
9. Foundations made from a large slab of concrete (suited to warmer climates).
10. The strong piece of wood nailed across the top of a window or door opening.
12. The hot water and cold water supplies and drainage system that are separate from each other.
13. Plywood, particle board, wooden planks, or rigid foam board that protects a house from weather.
14. Provides the surface to which the finished floor is attached.

232

CHAPTER 12
MANAGING PRODUCTION SYSTEMS

ACROSS

3. One who buys large quantities of products from manufacturers and resells this to customers.
5. One who designs the major systems of a building, such as heating and electrical systems.
9. A company hired by a general contractor to do plumbing, carpentry, or masonry work.
15. A product that is not repairable after the warranty period is up.
16. A test version of a product.
17. One who designs a building's shape and chooses materials for it.
18. A percentage of the sales price paid to the sales representative.
19. Person or company who takes the overall responsibility for a construction project.

DOWN

1. Changing a product to make it easier to manufacture and add features that are needed.

2. Written agreement between the owner and the general contractor.
3. One who sells products to people by taking orders and placing these orders with a manufacturer.
4. When more income comes in than money goes out.
6. This analysis predicts how much money will have to be spent each week or month.
7. _____ managers oversee contracts, scheduling, material deliveries, and progress of jobs.
8. A _____ of Occupancy must be issued before a building can be occupied.
10. These are necessary to protect both workers and people who pass by or work at construction sites.
11. A scheduling tool which lists each step necessary to complete a project.
12. A person who buys stock in a company.
13. Additional costs of a project caused by poor management or changes requested by the owner.
14. A bar graph that shows when a particular kind of work is to be started and finished.

CHAPTER 13
ENERGY

ACROSS

3. Mixture of gasoline and alcohol.
6. A curved reflector that focuses light to a single focal point.
7. Energy produced by the power from falling water.
12. Forcing of nuclei of two atoms together resulting in a release of a large amount of energy.
13. One quadrillion Btus, or the energy given off by burning 20 gallons of gas every day for a year.
14. The use of micro-organisms to turn biomass such as grain into alcohol and carbon dioxide gas.
15. Fuels such as oil, natural gas, and coal.
16. A photovoltaic cell which turns light into electricity.

DOWN

1. Processing of biomass in which methane gas is produced as the biomass rots.
2. Done when a force pushes or pulls on an object causing the object to move.
4. The ability to do work.
5. Vegetation and animal wastes.
8. Waste material of high-energy particles that can cause burns and sickness.
9. _____ of Energy: Energy cannot be created or destroyed, but only changed from one form to another.
10. The splitting of atomic nuclei resulting in other splitting and release of heat and light energy.
11. A system where the sun directly heats water.

CHAPTER 14
POWER

ACROSS

1. Amount of work done during a given time period.
4. A way to store energy in a power system that can be recharged.
7. Material that has no electrical resistance.
8. A type of engine that requires no spark plug to ignite the fuel and air.
10. A type of engine that carries its own liquid oxygen and does not require outside air to operate.
11. A machine that uses energy to create mechanical force and motion.
14. The non-moving housing in a generator.
15. Transmission system that uses water or other liquids to transmit force.
18. Temperature at which molecules stop moving.
19. Type of current produced by generators at power plants.

DOWN

1. The force on a liquid or object divided by the area over which it is applied.
2. A type of engine in which the fuel is also the expanding gas which moves the piston.
3. Used to change the high voltage to lower voltage near the place where electricity is used.
5. Carries force from one place to another or changes its direction.
6. The measurement of power, one of which equals 550 foot-pounds per second.
9. Both jets and rockets have _____ engines.
12. When a piston makes the object or _____ move, work is done.
13. The type of reactor in which the nuclei of an atom are split to create energy.
16. A type of engine in which air is pushed into a combustion engine by a compressor.
17. Measure of power equal to one kilogram-meter per second.

CHAPTER 15
TRANSPORTATION

ACROSS

3. An upward force exerted on an object by a fluid.
8. Similar to a gasoline engine, except it has no spark plugs.
9. A ship used to carry trailer-truck-size boxes.
11. This type of engine forces air into a combustion chamber, mixes the air with a fuel spray, and ignites the mixture.
12. This device can power craft in places where there is no oxygen since it carries its own supply.
13. To travel on a regular basis from home to work.
15. The gears or belts that connect a motor to the wheels.
17. Transportation systems that use different modes for different parts of a trip.
18. A type of external combustion engine used in early vehicles.

DOWN

1. The type of engine most often used to power automobiles.
2. A non-vehicle transportation system used to move crude oil or natural gas.
4. This propels an airplane or a ship.
5. The forward force (produced by an engine) that moves a vehicle.
6. A container that carries people or cargo.
7. An aircraft engine in which a turbine drives a propeller.
10. A non-vehicle transportation system used to move parts from one work station to another.
14. The upward force that must be created to get an airplane to fly.
16. The wind resistance that tends to hold back an airplane when it moves forward.

CHAPTER 16
IMPACTS FOR TODAY AND TOMORROW

ACROSS

3. Before implementing a technology, we must consider its effects on this.
5. Buildings with data communications lines in addition to electric wiring.
7. This futuristic communication technology will permit us to talk to a typewriter.
10. This technology allows us to hold face-to-face meetings with people in distant places.
11. The flowing together of technological systems.
13. Not just a modern-day problem.
14. Personal rapid transport (abbreviation.)
15. A plant that can be irrigated with salt water.
16. Techniques for forecasting the future.

DOWN

1. A futures _____ forecasts the possible outcomes of the event at the center.
2. A clean, safe, atomic energy source for the future.
4. Maglev.
6. Artificial speech produced by a computer.
8. Human factors engineering that fits technology to human needs.
9. Pollution from automobile exhaust and burning coal causes this.
12. Effect of a technological system on people or the environment.

ANSWERS TO CROSSTECH PUZZLES

Chapter 1 Technology in a Changing World

Chapter 2 Resources for Technology

Chapter 3 Problem Solving and Systems

Chapter 4 The Electronic Computer Age

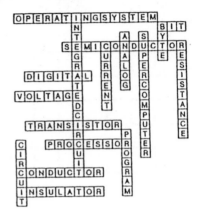

Chapter 5 Communication Systems

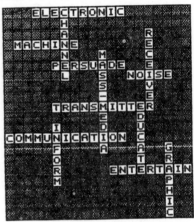

Chapter 6 Graphic Communications

Chapter 7 Electronic Communication

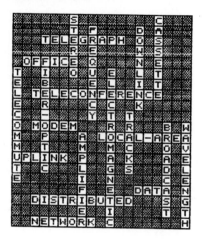

Chapter 8 Processing Materials

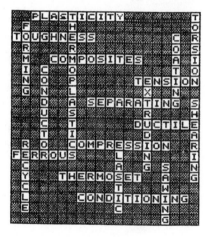

Chapter 9 Manufacturing

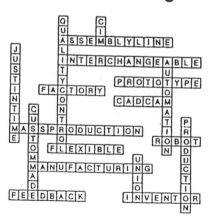

Chapter 10 Construction

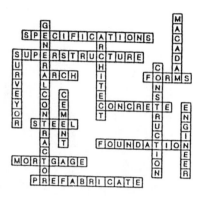

Chapter 11 Building a Structure

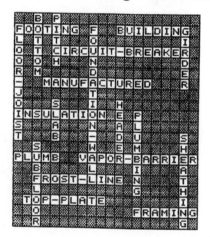

Chapter 12 Managing Production Systems

Chapter 13 Energy

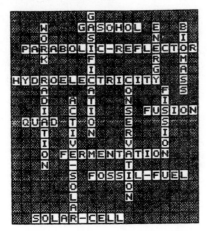

Chapter 15 Transportation

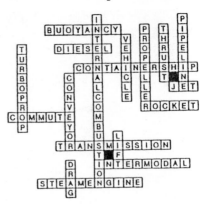

Chapter 14 Power

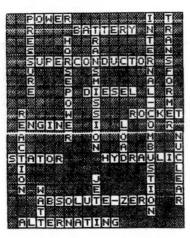

Chapter 16 Impacts for Today and Tomorrow

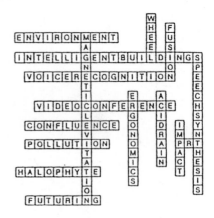

APPENDIX D
CHAPTER EVALUATION MATERIAL

TECHNOLOGY IN A CHANGING WORLD

Chapter 1, Text pages 2–25

KEY WORDS

Using the Key Words listed below, demonstrate your mastery of the Key Words by selecting the correct answer to complete each statement.

Agricultural
Bronze Age
Change
Communications
Construction
Energy
Exponential

Industrial Revolution
Information Age
Iron Age
Manufacturing
Mass production
Resources
Routines

Science
Stone Age
Technology
Technologically
   literate

1. During the _____ _____ people made tools and weapons from a mixture of copper and tin.

2. During the _____ era, people lived off of the land.

3. The rate of change of technology during the Information Age is described as _____.

4. Life is made up of _____. This is the way to go about doing things on a regular basis each day.

5. In the _____ _____ , inventions are based on electronics and the computer.

6. _____ is the study of why natural things happen the way they do.

7. Bones, wood, and stone were used to make tools during the _____ _____.

8. With the invention of machines came the _____ _____ .

9. The use of knowledge to turn resources into goods and services that society needs is called _____.

10. Using an assembly line to produce items on a large scale is called _____ _____.

TECHNOLOGY IN A CHANGING WORLD

Chapter 1, Text pages 2–25

REVIEW

MULTIPLE CHOICE: Place the letter of the correct answer in the blank provided for each question.

_____ 1. Routines can be influenced by:
   a.   the solar cycle.
   b.   technology.
   c.   both the solar cycle and technology.

_____ 2. The study of why natural things happen the way they do is called:
   a.   science.
   b.   routines.
   c.   technology.

_____ 3. The use of knowledge to turn resources into the goods and services that society needs is called:
   a.   routines.
   b.   technology.
   c.   science.

_____ 4. The schedule or procedure one goes through daily is called:
   a.   a technological discovery.
   b.   a routine.
   c.   scientific discovery.

_____ 5. Richard Arkwright was the inventor of the:
   a.   water wheel
   b.   metal plow
   c.   factory system

FILL IN THE BLANKS: In the blanks provided, write the correct answer to each question.

6. Why should we study technology?

   _____

   _____

   _____

7. List people's basic needs and wants that technology satisfies:

   a. _____   d. _____

   b. _____   e. _____

   c. _____   f. _____

8. List and describe the three distinct (not the historical) periods of technology:

    a. _____   _____

    _____.

    b. _____   _____

    _____.

    c. _____   _____

    _____.

TRUE/FALSE: Write the full word TRUE or FALSE in response to the following statements:

_____ 9. Human beings have been creating and using technology since the Industrial Revolution.

_____ 10. Henry Ford is known as the father of mass production.

_____ 11. The exponential rate of change of technology refers to the ever-increasing rate of change.

_____ 12. During the Agricultural Era, tools and discoveries were related to improving methods of tilling the soil.

_____ 13. During the Bronze Age and the Iron Age, tools and devices were made entirely from stones strapped to sticks.

_____ 14. The factory system, which began during the Industrial Revolution, produced goods quickly, but these goods cost more to produce.

_____ 15. The water wheel started the machine age.

CREATING OUR OWN WORLD

Chapter 1, Text pages 2–25

INTERFACING WITH TECHNOLOGY

Technology is the use of knowledge to turn resources into goods and services that society needs. To be prepared to do this, one must prepare for the use of technology. We can better prepare ourselves through a study of a wide variety of subjects.

SCIENCE: How does science relate to technology?

_____

_____

MATH: Mathematically explain what is meant by the exponential rate of growth of technology in the past years:

_____

_____

SOCIAL STUDIES: How does knowing about our past affect present and future technology?

_____

_____

LANGUAGE ARTS: What role does language arts play in our ability to function in a technological world?

_____

_____

ART: Why would one need artistic skills and knowledge to work in a technological field?

_____

_____

NAME _____ SCHOOL _____

CLASS _____ PERIOD _____ DATE _____ SCORE _____

TECHNOLOGY IN A CHANGING WORLD

Chapter 1, Text pages 2–25

TEST

TRUE/FALSE: Write the full word TRUE or FALSE in response to the following statements.

_____ 1. Science is the application of math and technology.

_____ 2. During the Agricultural Era, tools and discoveries were related to improving methods of tilling the soil.

_____ 3. The invention of machines occurred during the Stone Age.

_____ 4. The rate of change of technology during the Information Age is described as exponential.

_____ 5. Routines are influenced by technology.

FILL IN THE BLANKS: In the blanks provided, write the correct answer to each question.

6. During the _____ _____ people made tools and weapons from a mixture of copper and tin.

7. Using an assembly line to produce items on a large scale is called _____ _____.

8. The use of knowledge to turn resources into the goods and services that society needs is called _____.

9. Technology satisfies people's basic _____ and wants.

10. Richard Arkwright was the inventor of the _____ system.

TECHNOLOGY IN A CHANGING WORLD

Chapter 1, Text pages 2–25

ANSWER SHEET

////////// KEY WORDS \\\\\\\\\\

1. Bronze Age
2. Agricultural Era
3. Exponential
4. Routines
5. Information Age
6. Science
7. Stone Age
8. Industrial Revolution
9. Technology
10. Mass production

////////// REVIEW \\\\\\\\\\

MULTIPLE CHOICE:

1. c   2. a   3. b   4. b   5. c

FILL IN THE BLANKS:

6. People should study technology because we have come to depend on it to fulfill our basic needs and to make our lives more comfortable, healthy, and productive. (Answers may vary. Consult text for other responses.)

7. a. produce food
   b. medical needs
   c. manufacture items
   d. energy sources
   e. communicate ideas
   f. transportation needs

8. a. Agricultural — people lived off the land
   b. Industrial — many mechanical devices were invented
   c. Information — time when inventions are based on electronics and the computer

TRUE/FALSE:

9. False   10. False   11. True   12. True   13. False   14. False   15. True

////////// INTERFACING WITH TECHNOLOGY \\\\\\\\\\

SCIENCE: Technology is the use of what is learned about science. An example of this would be that we study science to learn about the sun, planets, and stars so that we can build vehicles to travel into outer space.

MATH: Technology has grown at an ever-increasing rate. We learn and invent more and more as each year passes.

SOCIAL STUDIES: Knowing about our past provides us with information that can be applied to decisions we make about future invention and use of technologies.

LANGUAGE ARTS: One must be able to read and write to be able to write letters, vote, and speak out about what is happening in technology.

ART: With the many technologies used to create designs using shapes, colors, and textures, we need to understand the impact of design on our decision to choose among products.

////////// TEST \\\\\\\\\\

TRUE/FALSE:

1. False
2. True
3. False
4. True
5. True

FILL IN THE BLANKS:

6. Bronze Age
7. mass production
8. technology
9. needs
10. factory

RESOURCES FOR TECHNOLOGY

Chapter 2, Text pages 26–43

KEY WORDS

Using the Key Words listed below, demonstrate your mastery of the Key Words by selecting the correct answer to complete each statement.

Capital	Inclined plane	People
Coal	Information	Pesources
Energy	Laser	Solar
Finite	Machines	Synthetic
Gas	Material	Time
Geothermal	Nuclear energy	Tools
Hydroelectricity	Oil	

1. Technology has been used to create substitutes for resources that are called _____.

2. The inclined plane, lever, pulley, wheel and axle, screw, and wedge are known as simple _____.

3. A/An _____ is the result of bursts of light that create light energy.

4. _____ energy is created by heat from inside the earth.

5. Any form of wealth such as stock, cash, buildings, machinery, and land, is known as _____.

6. _____ is a resource that comes from raw data that is processed by collecting, recording, calculating, classifying, storing, and retrieving it.

7. Energy resources formed from decayed plant and animal matter include _____, _____, and _____.

8. Energy from the sun is known as _____ energy.

9. Natural resources are _____ that are found in nature.

10. Technology comes from _____, a resource that creates and uses the technologies.

RESOURCES FOR TECHNOLOGY

Chapter 2, Text pages 26-43

REVIEW

FILL IN THE BLANKS: In the blanks provided, write the correct answer to each question.

1. List and explain the seven types of resources used in technological systems:

   a. _____

   b. _____

   c. _____

   d. _____

   e. _____

   f. _____

   g. _____

2. Explain the difference between these two raw materials:

   a. Renewable _____

   b. Nonrenewable _____

3. List the six energy categories:

   a. _____   d. _____

   b. _____   e. _____

   c. _____   f. _____

4. Synthetic materials are ones that are _____

   _____

5. Capital resources include any form of wealth such as:

   a. _____   d. _____

   b. _____   e. _____

   c. _____

MATCHING: Place the letter of the correct response in the blank to the left of the term it matches.

_____ 6. Solar energy
_____ 7. Natural resources
_____ 8. Synthetic
_____ 9. Geothermal
_____ 10. Laser
_____ 11. Renewable
_____ 12. Nonrenewable
_____ 13. Chemical energy
_____ 14. Information
_____ 15. Machines

A. energy from heat inside the earth
B. change the amount, speed, or direction of a force
C. produces wind, heat, and light
D. transmits bursts of light energy
E. comes from wood and fossil fuels
F. materials found in nature
G. resource found in computer files
H. materials that can be replaced
I. materials made in a laboratory
J. materials that cannot be replaced

IDENTIFICATION: Identify the following simple machines.

16. _____

17. _____

18. _____

19. _____

20. _____

21. _____

RESOURCES FOR TECHNOLOGY

Chapter 2, Text pages 26–43

INTERFACING WITH TECHNOLOGY

Technological systems are made up of seven types of resources: people, information, materials, tools and machines, energy, capital, and time. The resources can be classified as limited and unlimited or renewable and nonrenewable. To use resources wisely, we need to become well informed.

SCIENCE: What is studied in science concerning resources?

_____

_____

MATH: How does our ability to understand and work with math affect our decisions regarding the use of our resources?

_____

_____

SOCIAL STUDIES: What do we learn in social studies about the use of resources such as machinery?

_____

_____

LANGUAGE ARTS: Why do we need to be able to read and comprehend information relating to our resources?

_____

_____

ART: What resources are used in the creation of art work?

_____

_____

RESOURCES FOR TECHNOLOGY

Chapter 2, Text pages 26–43

TEST

TRUE/FALSE: Write the full word TRUE or FALSE in response to the following statements.

_____ 1. Coal, oil, and gas are examples of nonrenewable resources.

_____ 2. Synthetic materials are grown in nature and can be replaced.

_____ 3. A laser produces light energy that is used to measure, cut, weld, and communicate messages.

_____ 4. Information is a resource found in computer files, books, and magazines.

_____ 5. Capital is a resource such as computer files, books, and magazines.

MATCHING: Place the letter of the correct answer in the blank provided for each question.

_____ 6. Gravitational

_____ 7. Limited energy

_____ 8. Machines

_____ 9. Geothermal

_____ 10. Renewable

A. comes from heat deep in the earth

B. energy from tides and falling water

C. grown or replaceable form of energy

D. lever, screw, pulley, and wedge

E. energy such as coal, oil, and gas

FILL IN THE BLANKS: In the blanks provided, write the correct answer to each question.

11. The seven resources used in technological systems are:

a. _____ e. _____

b. _____ f. _____

c. _____ g. _____

d. _____

12. Five forms of capital are:

a. _____ d. _____

b. _____ e. _____

c. _____

RESOURCES FOR TECHNOLOGY

Chapter 2, Text pages 26–43

ANSWER SHEET

///////// **KEY WORDS** \\\\\\\\\

1. Synthetics
2. Machines
3. Laser
4. Geothermal
5. Capital
6. Information
7. Coal, oil, and gas
8. Solar
9. Materials
10. People

///////// **REVIEW** \\\\\\\\\

FILL IN THE BLANKS:

1. (Student answers may vary)
   a. People . . . design, create, and use technology.
   b. Information . . . is data that are processed.
   c. Materials . . . are either raw materials or synthetics.
   d. Tools & Machines . . .can be hand, machine, electronic, or optical.
   e. Energy . . . is classified into human and animal muscle power, chemical, solar, geothermal, gravitational, and nuclear.
   f. Capital . . . is any form of wealth such as stock, cash, pieces of equipment, buildings, and land.
   g. Time . . . is measurement of a period or duration in increments as small as nanoseconds (billionths of a second.)

2. a. Renewable materials are those that can be grown and therefore replaced.
   b. Nonrenewable materials cannot be grown or replaced.

3. a. Human and animal muscle   c. Chemical   e. Geothermal
   b. Solar   d. Gravitational   f. Nuclear

4. Synthetic materials are ones that are made in a laboratory.

5. a. Cash   c. Buildings   e. Land
   b. Shares of Stock   d. Pieces of equipment

MATCHING:

6. C   8. I   10. D   12. J   14. G
7. F   9. A   11. H   13. E   15. B

IDENTIFICATION:

16. Wedge   18. Screw   20. Pulley
17. Inclined plane   19. Lever   21. Wheel and axle

///////// **INTERFACING WITH TECHNOLOGY** \\\\\\\\\

(Answers may vary and there may be additions to these suggested answers. Answers can depend upon the level of studies in the various disciplines for the group of students you are teaching.)

SCIENCE: In science we study the solar system and its effects on the earth in the development of our natural resources. We study forms of energy and how these are converted into products. We learn about the composition of synthetics.

MATH: When we use resources, we must be aware of the amounts available, how much we are using, and how much this costs. Knowing this affects our use of technological resources.

SOCIAL STUDIES: Studying our past develops in us an awareness of the progression of technological developments. It helps us to understand how fast technology is advancing and how we need to prepare for this.

LANGUAGE ARTS: We need to be able to read literature such as books, newspapers, brochures, and magazines to wisely make choices in determining the use of our natural resources.

ART: Materials found in nature such as clay, pigments, and elements are a basis for the media used in developing art work.

///////// TEST \\\\\\\\\

TRUE/FALSE:

1. True   2. False   3. True   4. True   5. False

MATCHING:

6. B   7. E   8. D   9. A   10. C

FILL IN THE BLANKS:

11. a. People   d. Capital   f. Time
    b. Information   e. Energy   g. Materials
    c. Tools and machines

12. a. Cash   d. Buildings
    b. Shares of stock e. Pieces of equipment
    c. Land

SYSTEMS AND PROBLEM SOLVING

Chapter 3, Text pages 44–86

KEY WORDS

Using the Key Words listed below, demonstrate your mastery of the Key Words by selecting the correct answer to complete each statement.

Actual results	Feedback	Output
Alternatives	Implementation	Problem solving
Basic research	Input	Process
Basic systems model	Insight	Prototype
Brainstorming	Market research	Research
Closed-loop system	Model	Subsystem
Control	Monitoring	System
Design brief	Open-loop system	Trade-off
Design folder	Optimization	Trial and error
Desired results		

1. The action part of the system is the _____.

2. Information about the output is called _____.

3. What is produced or the actual result is the _____ .

4. Systems that have feedback are called _____ _____.

5. The smaller systems that make up a larger system are called _____.

6. _____ is coming up with different solutions to solve a problem.

7. Being thorough in researching a problem and creative in thinking is the use of _____ to develop alternatives in finding solutions.

8. Selection of a solution, even though it is not perfect, is called making a _____.

9. Information gathered for the purpose of problem solving is called _____ _____.

10. A _____ is a full-sized or smaller-scale version of a proposed solution to a problem.

SYSTEMS AND PROBLEM SOLVING

Chapter 3, Text pages 44-86

REVIEW

MULTIPLE CHOICE: Place the letter of the correct answer in the blank provided for each question.

_____ 1. A means of getting a desired result is called:
   a. a system
   b. output
   c. input
   d. a subsystem

_____ 2. The action part of a technological system or how the system will achieve the desired result is the:
   a. system
   b. output
   c. input
   d. process

_____ 3. The command we give a system is the:
   a. output
   b. process
   c. input
   d. monitor

_____ 4. The actual results of a system that the process produces are called the:
   a. output
   b. process
   c. input
   d. feedback

_____ 5. Information about the output of a system is called:
   a. feedback
   b. inputs
   c. control systems
   d. outputs

_____ 6. Systems that have feedback are called:
   a. open-loop systems
   b. double-loop systems
   c. single-loop systems
   d. closed-loop systems

_____ 7. Feedback about an output can be done by a/an:
   a. subsystem
   b. monitor
   c. input
   d. output

_____ 8. Systems can be made up of many smaller systems called:
   a. inputs
   b. outputs
   c. subsystems
   d. processes

FILL IN THE BLANKS: In the blanks provided, write the correct answer to each question.

9. Four types of output include:

   a. _____   c. _____

   b. _____   d. _____

10. The seven steps of the technological methods of problem solving involve:

    a. _____

    b. _____

    c. _____

    d. _____

    e. _____

    f. _____

    g. _____

11. List five concerns in solving real-world problems:

    a. _____

    b. _____

    c. _____

    d. _____

    e. _____

12. Five types of models include:

    a. _____   d. _____

    b. _____   e. _____

    c. _____

IDENTIFICATION: IDENTIFY THE PARTS OF THIS SYSTEM MODEL:

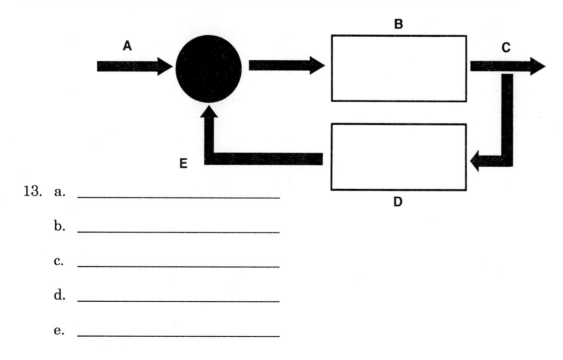

13. a. _____

    b. _____

    c. _____

    d. _____

    e. _____

SYSTEMS AND PROBLEM SOLVING

Chapter 3, Text pages 44–86

INTERFACING WITH TECHNOLOGY

Problem solving is a step-by-step procedure used find solutions. A system is a means of getting a desired result. Technological systems are all alike in that each has an input, process, and output. Knowing about problem solving and systems requires knowledge of other disciplines. Below, tell how other subjects you study relate to your studies in technology.

SCIENCE: By what process are problems solved in the scientific field?

_____

_____

MATH: How do you think math is used in problem solving?

_____

_____

SOCIAL STUDIES: What do we learn in social studies regarding the history of development of products we use?

_____

_____

LANGUAGE ARTS: How does our ability to read and write impact our ability to understand and go about the problem-solving process?

_____

_____

ART: How is art used in the problem-solving process?

_____

_____

SYSTEMS AND PROBLEM SOLVING

Chapter 3, Text pages 44–86

TEST

TRUE/FALSE: Write the full word TRUE or FALSE in response to the following statements.

_____ 1. The actual results of a system that the process produces are called the output.

_____ 2. The action part of a technological system or how the system will achieve the desired result is the process.

_____ 3. Feedback is information about the input.

_____ 4. Systems that have feedback are called closed-loop systems.

_____ 5. A means of achieving a desired result is called monitoring.

MULTIPLE CHOICE: Place the letter of the correct answer in the blank provided for each question.

_____ 6. The outputs of a system can be expected and desirable, or the outputs can be:
  a. exponential   c. unexpected
  b. undesirable   d. Both (b) and (c)

_____ 7. The first step of problem solving is:
  a. setting goals   c. defining the problem
  b. evaluation   d. alternative solutions

_____ 8. Coming up with different ideas in problem solving is called:
  a. setting goals   c. defining the problem
  b. evaluation   d. developing alternatives

_____ 9. The final step in problem solving is:
  a. implementation   c. defining the problem
  b. evaluation   d. developing alternatives

_____10. The command we give a system is the:
  a. output   c. input
  b. process   d. feedback

SYSTEMS AND PROBLEM SOLVING

Chapter 3, Text pages 44–86

ANSWER SHEET

/////////// KEY TERMS \\\\\\\\\\\

1. Process
2. Feedback
3. Output
4. Closed-loop systems

5. Subsystems
6. Research
7. Insight

8. Trade-off
9. Market research
10. Model

/////////// REVIEW \\\\\\\\\\\

MULTIPLE CHOICE:

1. a   2. d   3. c   4. a   5. a   6. d   7. b   8. c

FILL IN THE BLANKS:

9. a. expected, desirable
    b. expected, undesirable
    c. unexpected, desirable
    d. unexpected, undesirable

10. a. Describe the problem
    b. Describe the results you want
    c. Gather information
    d. Think of alternative solutions
    e. Choose the best solution
    f. Implement the solution
    g. Evaluate the solution and make necessary changes

11. a. social and environmental concerns
    b. politics
    c. risk/benefit trade-offs
    d. need of continued monitoring
    e. values – also culture

12. a. charts and graphs
    b. mathematical models
    c. sketches, illustrations, and technical drawings
    d. working models
    e. computer simulation

13. a. INPUT
    b. PROCESS
    c. OUTPUT
    d. MONITOR
    e. FEEDBACK

/////////// INTERFACING WITH TECHNOLOGY \\\\\\\\\\\

(Answers may vary and there may be additions to these suggested answers. Answers can depend upon the level of studies in the various disciplines for the group of students you are teaching.)

SCIENCE: The scientific method is used to solve problems in science.

MATH: Part of the problem-solving process may need calculations of measurements and quantities.

SOCIAL STUDIES: In social studies, we learn of the past achievements in technology, the successes and failures, and the impacts upon the history of a nation.

LANGUAGE ARTS: To be able to solve problems, we must be able to read information to form a basis for development of possible solutions. We must also be able to express in an acceptable manner the process we follow and the resulting solutions.

ART: In problem solving, we need to be able to sketch ideas. Art will give us the tools to design and sketch solutions to problems.

////////// **TEST** \\\\\\\\\\

TRUE/FALSE	**MULTIPLE CHOICE:**
1. true	6. d
2. false	7. c
3. false	8. d
4. true	9. b
5. false	10. c

THE ELECTRONIC COMPUTER AGE

Chapter 4, Text pages 88–119

KEY WORDS

Using the Key Words listed below, demonstrate your mastery of the Key Words by selecting the correct answer to complete each statement.

Ampere	Conductor	Memory	Random access
Analog	Current	Operating system	memory (RAM)
Bit	Digital	Printed circuit	Resistance
Byte	Electron	board	Semiconductor
Circuit	I/O	Printer	Supercomputer
Component	Insulator	Processor	Transistor
Compounds	Integrated circuit	Program	Voltage

1. The flow of electrons is called the _____.

2. A/An _____ is a material that is neither a good insulator nor a good conductor.

3. Groups of components grouped together to do a specific job are called _____.

4. A complete electronic circuit made at one time on a piece of semiconductor material is known as a/an _____ _____.

5. In a/an _____ circuit, information is relayed by changes in voltage.

6. A computer does its work using a set of instructions called a_____.

7. The place where the program is stored is called the _____.

8. The _____ _____ _____ stores the program and information that is being worked on.

9. Information provided through the use of a keyboard is called the _____.

10. A _____ is a computer output device that records output on paper.

THE ELECTRONIC COMPUTER AGE

Chapter 4, Text pages 88–119

REVIEW

MULTIPLE CHOICE: Place the letter of the correct answer in the blank provided to the left of each question.

_____ 1. Materials that are made up of atoms of only one type are called:
    a. neutrons   c. protons
    b. elements   d. electrons

_____ 2. Materials whose atoms give up some electrons easily are called:
    a. elements   c. conductors
    b. insulators   d. both "B" and "C"

_____ 3. The measurement of electric current flow is the:
    a. voltage   c. ampere
    b. resistance   d. conductor

_____ 4. The pressure applied to electrons in order to get current to flow is called:
    a. electromotive force   c. resistance
    b. amperage   d. insulation

_____ 5. The unit of measurement of opposition to the flow of current is the:
    a. volt   c. amp
    b. insulator   d. ohm

FILL IN THE BLANKS: In the blanks provided, write the correct answer to each question.

6. _____ control the flow of electricity and perform useful tasks.

7. _____ are groups of components connected together to perform a specific function.

8. _____ are materials that are neither good insulators nor good conductors. Usually made of silicon; a diode is an example of this.

9. A/an _____ allows a small amount of current to control the flow of a much larger amount of current.

10. A/an _____ is a resistor whose resistance changes with temperature.

11. A/An _____ _____ _____ is usually made of fiberglass with a thin sheet of copper that has patterns etched in it to form conducting paths for current.

12. A/an _____ _____ provides a complete circuit function on a single piece of semiconductor material.

13. A _____ is a photographic reduction of a chip drawn several hundred times larger than its actual circuit.

14. In _____ circuits, voltages change very smoothly.

15. In _____ circuits, information is first coded into a series of 0s and 1s, each called a bit (short for binary digit).

MATCHING: The following terms refer to computers. Place the letter of the correct response in the blank to the left of the term it matches.

_____16. Program
_____17. Memory
_____18. Input devices
_____19. Output
_____20. Microcomputers
_____21. Modem
_____22. MS-DOS
_____23. BASIC
_____24. Pascal
_____25. Artificial intelligence

A. imitation of human thought by computers
B. small computers found in appliances and automobiles
C. list of instructions that tells a computer how to perform its work
D. Beginner's All-purpose Symbolic Instruction Code
E. place where information being worked on is stored and kept
F. mouse, light pen, and the touch sensitive screen
G. one type of operating system of a personal computer
H. forms of this include video monitors and hard copies
I. general-purpose programming language named after a mathematician
J. sends computer data over telephone lines

TRUE/FALSE: Write the full word TRUE or FALSE in response to the following statements.

_____ 26. Transistors, diodes, resistors, conducting paths, and other circuit components are made on the computer chip.

_____ 27. Charles Babbage invented the first mechanical adding machine.

_____ 28. Secondary storage in a computer includes floppy disks, hard disks, and magnetic tape.

_____ 29. Mainframe computers are the fastest computers and are used for analyzing satellite data.

_____ 30. GFLOPS are programs that are not written correctly.

THE ELECTRONIC COMPUTER AGE

Chapter 4, Text pages 88–119

INTERFACING WITH TECHNOLOGY

Electronics has changed technology significantly. Computers are general-purpose tools of the Information Age. They are used in appliances, automobiles, word processing, science research, manufacturing, and business. Coexisting with computers requires knowledge of other disciplines. Below you will be asked to give input concerning the relationship of studying electronics and computers in technology to the other disciplines.

SCIENCE: What have you studied in science that relates to electronics?

_____

_____

MATH: How does math fit into a study of computers and how they are programmed?

_____

_____

SOCIAL STUDIES: What is learned in history that relates to computers?

_____

_____

LANGUAGE ARTS: How does our ability to read and write help us when working with electronics and computer technology?

_____

_____

ART: How can art be helpful when using a computer?

_____

_____

NAME_____ SCHOOL_____

CLASS_____ PERIOD_____ DATE_____ SCORE_____

THE ELECTRONIC COMPUTER AGE
Chapter 4, Text pages 88–119

TEST

TRUE/FALSE: Write the full word TRUE or FALSE in response to the following statements.

_____ 1. Materials whose atoms give up electrons easily are called conductors.

_____ 2. Output devices include the light pen, keyboard, and the touch-sensitive screen.

_____ 3. Mainframe computers are the fastest computers and are used for analyzing satellite data.

_____ 4. A thermistor is a resistor whose resistance changes with the temperature.

_____ 5. Electromotive force is the measurement of electrical impulses in a printed circuit.

_____ 6. MS-DOS is one type of operating system for a computer.

_____ 7. Transistors, diodes, resistors, and conducting paths are all made on the chip.

_____ 8. Charles Babbage invented the first mechanical adding machine.

_____ 9. GFLOPS programs are ones that are not written correctly.

_____ 10. The measurement of electric current is the amp.

MATCHING: Place the letter of the correct response in the blank to the left of the term it matches.

____11. BASIC
____12. Pascal
____13. Modem
____14. Program
____15. Semiconductor
____16. Ohm
____17. Synthesizer
____18. Hard disk
____19. Circuits
____20. Artificial intelligence

A. sends computer data over telephone lines
B. a type of secondary storage system in a computer
C. imitation of human thought by computers
D. device that generates music inside a computer
E. a measurement of resistance to the flow of electric current
F. general-purpose programming language named after a mathematician
G. groups of components joined together to perform a specific function
H. list of instructions that tells a computer how to perform its work
I. Beginners All-purpose Symbolic Instruction Code
J. materials usually made of silicon that are neither good nor bad conductors

THE ELECTRONIC COMPUTER AGE

Chapter 4, Text pages 88–119

ANSWER SHEET

//////// KEY TERMS \\\\\\\\

1. current
2. semiconductor
3. circuits
4. printed circuit
5. analog
6. program
7. memory
8. random access memory
9. input
10. printer

//////// REVIEW \\\\\\\\

MULTIPLE CHOICE:

1. b
2. c
3. c
4. a
5. d

FILL IN THE BLANKS:

6. Electronic components
7. Circuits
8. Semiconductors
9. transistor
10. thermistor
11. printed circuit board
12. integrated circuit
13. mask
14. analog
15. digital

MATCHING:

16. C
17. E
18. F
19. H
20. B
21. J
22. G
23. D
24. I
25. A

TRUE/FALSE:

26. True
27. False
28. True
29. False
30. False

//////// INTERFACING WITH TECHNOLOGY \\\\\\\\

(Answers may vary and there may be additions to these suggested answers. Answers can depend upon the level of studies in the various disciplines for the group of students you are teaching.)

SCIENCE: In science, you study atoms, elements, compounds, and the flow of electrons. You learn how the ability to conduct, or to resist, the flow of electricity makes materials conductors and non-conductors, and how this is used to build components for electronic devices.

MATH: Most of today's computers are digital. To understand the concept behind programming a computer, one must understand the binary number system.

SOCIAL STUDIES: In history, you study about the development of devices for purposes of calculating such as the abacus, Napier's bones, calculators, and computers.

LANGUAGE ARTS: We need to be able to read and write in order to read instruction and repair manuals. We need to be able to express ourselves when we use Information Age devices.

ART: Our present high-tech computers require a knowledge of art elements when developing graphic presentations for reports, magazines, texts, and brochures.

//////// **TEST** \\\\\\\\

TRUE/FALSE:

1. True
2. False
3. False
4. True
5. False
6. True
7. True
8. False
9. False
10. True

MATCHING:

11. I
12. F
13. A
14. H
15. J
16. E
17. D
18. B
19. G
20. C

COMMUNICATION SYSTEMS

Chapter 5, Text pages 120-135

KEY WORDS

Using the Key Words listed below, demonstrate your mastery of the Key Words by selecting the correct answer to complete each statement.

Channel
Communication system
Educate
Electronic communication
Entertain

Graphic communication
Inform
Machine communication
Mass media

Noise
Persuade
Receiver
Transmitter

1. When people speak to each other, they create a _____ _____.

2. The purposes of a communication system are to _____, _____, and _____.

3. Radios, televisions, and newspapers are examples of _____ _____.

4. The channel carries pictures or printed words in _____ _____.

5. The channel carries electronic signals in _____ _____.

6. The _____ is the part of the communication process that sends the message.

7. The _____ is the part of the communication process that routes the message.

8. The _____ is the part of the communication process that accepts the message.

9. _____ is anything that interferes with communication.

10. _____ _____ occurs when machines send signals to each other for purposes of making something happen.

COMMUNICATION SYSTEMS

Chapter 5, Text pages 120–135

REVIEW

FILL IN THE BLANKS: In the blanks provided, write the correct answer to each question.

1. List and give an example of the three types of communication:

   a. _____   _____

   b. _____   _____

   c. _____   _____

2. Explain the meaning of these parts of a communication system:

   a. Input_____

   b. Process_____

   c. Output_____

   d. Feedback_____

3. List the three parts of the communication process and explain each:

   a. _____   _____

   b. _____   _____

   c. _____   _____

4. Communication systems are designed to reach a variety of goals. List these goals:

   a. _____

   b. _____

   c. _____

   d. _____

5. Two categories of communication are:

   a. _____   b. _____

DISCUSSION: Use complete sentences in forming your answer.

6. How does one choose a communication system? Give an example.

   _____

   _____

MATCHING: Below are the seven resources for communication technology. Match the examples of the resources to the correct resource.

_____ 7. People
_____ 8. Information
_____ 9. Materials
_____10. Tools & Machines
_____11. Energy
_____12. Capital
_____13. Time

A. depends on length and rate of speed
B. paper, film, and tape
C. money, equipment, and facilities
D. camera operators, actors, & writers
E. cameras, computers, & recorders
F. math, science, & practical knowledge
G. electricity

COMPLETION: Below is a chart showing the communication process. Fill in the boxes to give examples of each of the processes.

THE COMMUNICATION PROCESS

	TRANSMITTER	CHANNEL	RECEIVER
14. DATA COMMUNICATIONS			
15. MORSE CODE			
16. PHOTOGRAPHY			
17. PRINTING			
18. TYPING			
19. SPEECH			
20. WRITING LETTERS			

NAME_____ SCHOOL_____

CLASS_____ PERIOD_____ DATE_____ SCORE_____

COMMUNICATION SYSTEMS

Chapter 5, Text pages 120–135

INTERFACING WITH TECHNOLOGY

Communication includes having a message sent, received, and understood. To achieve this, a communication system requires the use of the seven technological resources. Humans, animals, and machines communicate. A knowledge of other disciplines enables us to understand and use communication systems. Below you will be asked to give input concerning the relationship of studying communications in technology to the other disciplines.

SCIENCE: What have you studied in science that relates to communication systems?

_____

_____

MATH: How could math benefit you in the understanding and use of communication systems?

_____

_____

SOCIAL STUDIES: What have you noted about communication systems in the past five years?

_____

_____

LANGUAGE ARTS: How does our ability to read and write help us when we are trying to communicate?

_____

_____

ART: Of what benefit is art to us when using a graphic communication system?

_____

_____

COMMUNICATION SYSTEMS

Chapter 5, Text pages 120–135

TEST

TRUE/FALSE: Write the full word TRUE or FALSE in response to the following statements concerning resources of communication systems.

_____ 1. Capital refers to time spent to rent or build work spaces, such as a TV studio or print shop.

_____ 2. People are the camera operators, computer operators, technicians, and set designers.

_____ 3. Information is the camera, computer, television studio, and print shop.

_____ 4. Energy moves the message from place to place.

_____ 5. Materials refer to the computers, cameras, printing presses, tape recorders, and printers.

_____ 6. Tools and machines are the paper, film, and tapes used in the communication system.

_____ 7. Time depends on the length of the message and the rate at which the message is sent.

MULTIPLE CHOICE: Place the letter of the correct answer in the blank provided for each question.

_____ 8. An example of machine-to-machine communication is:
    a. person on telephone   c. person speaking to pet
    b. computer to robot   d. person to computer

_____ 9. An example of animal communication is:
    a. person to dog   c. bee to bee
    b. person to computer   d. both "a" and "c"

_____10. In a communication system, the desired result is the:
    a. input   c. output
    b. process   d. feedback

_____11. In a communication system, how you will communicate is called the:
    a. input   c. process
    b. output   d. feedback

_____12. The message received by the other person is called the:
    a. output   c. feedback
    b. process   d. input

_____13. Whether or not someone has understood the message is called:
    a. output   c. feedback
    b. process   d. input

FILL IN THE BLANKS: In the blanks provided, write the answer to each statement.

14. _____ communication systems are those where the channel carries images or printed words.

15. _____ communication systems are those where the channel carries electrical signals.

16. _____ is anything that interferes with communication.

17. The purposes of communication are to _____, _____, _____, and _____.

COMMUNICATION SYSTEMS

Chapter 5, Text pages 120–135

ANSWER SHEET

///////// KEY WORDS \\\\\\\\\\\

1. communication system
2. educate, entertain, and inform
3. mass media
4. graphic communication
5. electronic communication

6. transmitter
7. channel
8. receiver
9. Noise
10. Machine communication

///////// REVIEW \\\\\\\\\\\

FILL IN THE BLANKS:

1. a. Person-to-Person — people communicating with each other
   b. Animal — people communicating with an animal or animal communicating with another animal (bees and chimpanzees)
   c. Machine — people communicating with machines (computers) or machines communicating with machines (computer and robot)

2. a. Input is the message you want to send to the other person.
   b. Process is how you will communicate the input.
   c. Output is the message the other person gets.
   d. Feedback tells you whether the person understood what you said.

3. a. Transmitter — send the message
   b. Channel — the route the message takes
   c. Receiver — that which accepts the message

4. a. Inform people about an event
   b. Persuade people to do something
   c. Entertain
   d. Educate people

5. a. electronic
   b. graphic

6. A communication system is chosen because it best satisfies a need. To inform people, you might choose a printed brochure. To persuade people, you might choose a telephone. You could cartoon pictures on television to entertain people. To educate, you might choose charts and drawings.

MATCHING:

7. D   8. F   9. B   10. E   11. G   12. C   13. A

COMPLETION:

14. a. Computer with monitor and modem   b. Telephone wires   c. Computer, modem

15. a. Morse code key and sending unit   b. Telegraph wires   c Morse code receiving unit

16. a. Light focused by a camera   b. Film   c. Snapshot

17. a. Printing press   b. Ink   c. Paper

18. a. Typewriter   b. Ribbon   c. Paper

19. a. Person   b. Air   c. Other person

20. a. Pen   b. Ink   c. Paper

///////// INTERFACING WITH TECHNOLOGY \\\\\\\\\

(Answers may vary and there may be additions to these suggested answers. Answers can depend upon the level of studies in the various disciplines for the group of students you are teaching.)

SCIENCE: In science, you learn about electricity and electromagnets. In technology you use the tools and machines run by electricity. You use printers that have electromagnets running the print-head.

MATH: Understanding math helps you to understand the tremendous speed (186,000 miles per second) electrical signals travel. This explains how we get messages so quickly.

SOCIAL STUDIES: In history, inventions and/or inventors of devices such as the printing press, computers, satellite communication and telephones are studied.

LANGUAGE ARTS: To be able to work with communication devices, one must be able to communicate and understand communications from others.

ART: In graphic communications, we need to understand and work with art concepts in designing the products of graphic communications such as posters, bumper stickers, advertisements, and billboards.

///////// TEST \\\\\\\\\

TRUE/FALSE:

1. False
2. True
3. False
4. True
5. False
6. False
7. True

MULTIPLE CHOICE:

8. B
9. D
10. A
11. C
12. A
13. C

FILL IN THE BLANKS:

14. Graphic
15. Electronic
16. Noise
17. entertain, educate, inform, and persuade

GRAPHIC COMMUNICATION

Chapter 6, Text pages 136–165

KEY WORDS

Using the Key Words listed below, demonstrate your mastery of the Key Words by selecting the correct answer to complete each statement.

CAD	Laser printer	Phototypesetter
Daisy wheel	Negative	Pictorial
Desktop publishing	Offset printing	Relief printing
Dot matrix	Orthographic	Screen printing
Gravure printing	Perspective	Technical drawing
Isometric		

1. A/An _____ drawing is drawn to show the object in three dimensions.

2. An important tool of engineers that does away with the use of a drawing board and T Square is _____.

3. A/An _____ drawing shows each side of an object in a separate picture.

4. _____ _____ is done using an ink which is pushed through a screen.

5. Metal sheets are used in _____ _____ which is how most printing is done.

6. A _____ _____ printer has a print head with many petals carrying the letters, numbers, and other printing characters at their tips.

7. A_____ _____ printer has a rectangular-shaped group of pins.

8. _____ _____ is also known as letterpress printing. Raised surfaces of the letters are inked.

9. _____ _____ is also known as intaglio printing. Scratches in the lower surface hold the ink.

10. A/An _____ drawing makes an object look natural. Parts of the object farther away are drawn smaller by the use of vanishing points on a horizon.

GRAPHIC COMMUNICATION

Chapter 6, Text pages 136–165

REVIEW

FILL IN THE BLANKS: In the blanks provided, write the correct answer to each question.

1. Examples of graphic communication systems include:

    a. _____ d. _____

    b. _____ e. _____

    c. _____ f. _____

2. List advantages of a CAD system over hand-drawn designs.

    a. _____

    b. _____

    c. _____

    d. _____

    e. _____

3. The five elements of photography are:

    a. _____ d. _____

    b. _____ e. _____

    c. _____

4. List the four types of cameras and briefly describe each.

    a. _____ _____

    b. _____ _____

    c. _____ _____

    d. _____ _____

5. Four types of printing are:

    a. _____ c. _____

    b. _____ d. _____

6. List three types of computer printers and briefly describe each.

  a. _____   _____

     _____

  b. _____   _____

     _____

  c. _____   _____

     _____

TRUE/FALSE: Write the full word TRUE or FALSE in response to the following statements.

_____ 7. Word processing combines typewriter and computer technologies.

_____ 8. An orthographic drawing is one that has one or more vanishing points.

_____ 9. CAD stands for computer-aided design or computer-aided drawing.

_____ 10. Johann Gottlieb invented the printing press.

_____ 11. Letterpress printing is the same as the gravure printing process.

_____ 12. A printer that produces high quality text and graphics very rapidly is the laser.

MULTIPLE CHOICE: Place the letter of the correct answer in the blank provided for each question.

____13. Styles of lettering are called:
  a. cursive   c. fonts
  b. CAD   d. sketches

____14. Photographic film is made out of:
  a. cotton paper   c. linen
  b. acetate   d. aluminum

____15. Copying machines make use of photography and:
  a. daisy wheels   c. static electricity
  b. offset presses   d. gravure printing

____16. Xerography means:
  a. dry writing   c. scanning
  b. typesetting   d. WYSIWYG

____17. All photography requires:
  a. a stencil   c. light
  b. printer's ink   d. CAD

IDENTIFICATION: Identify the following drawings of houses as perspective, isometric, oblique, or orthographic.

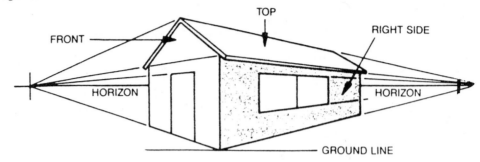

18. _____

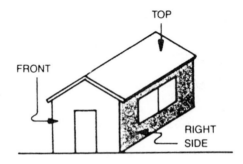

19. _____

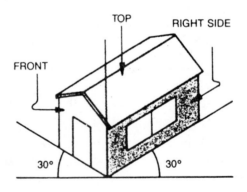

20. _____

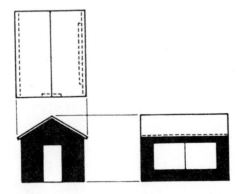

21. _____

GRAPHIC COMMUNICATION

Chapter 6, Text pages 136–165

INTERFACING WITH TECHNOLOGY

Graphic communications includes relief, gravure, screen, and offset printing. Word processing, desktop publishing, laser printing and photocopying are also forms of graphic communications. A knowledge of other disciplines enables us to understand and use graphic forms of communication. Below you will be asked to give input concerning the relationship of studying graphic communications in technology to the other disciplines.

SCIENCE: What do you learn in science that would help you if you were to work in the printing field?

_____

_____

MATH: How could math help you if you were to be a photographer and develop your own pictures?

_____

_____

SOCIAL STUDIES: How was history recorded in the past?

_____

_____

LANGUAGE ARTS: Why is language arts so important in the graphic communication field?

_____

_____

ART: How will art help you if you choose to go into advertising?

_____

_____

GRAPHIC COMMUNICATION

Chapter 6, Text pages 136–165

TEST

MULTIPLE CHOICE: Place the letter of the correct answer in the blank provided for each question.

_____ 1. Graphic communication systems include writing and:
a. technical drawing   c. photocopying
b. printing   d. "a," "b," and "c"

_____ 2. The computer printer that has a rectangular-shaped head made up of pins that strike a ribbon is called a:
a. laser   c. daisy wheel
b. dot matrix   d. obscura

_____ 3. The four types of printing include gravure, relief, screen, and:
a. offset   c. obscura
b. toner   d. Gothic

_____ 4. Xerography means:
a. zero defects   c. dry writing
b. zero mistakes   d. a zero graph

_____ 5. Photographic paper is made out of:
a. acetate   c. linen fibers
b. cotton fibers   d. aluminum

_____ 6. The printer that produces the highest quality pages is:
a. daisy wheel   c. laser
b. hand   d. dot matrix

_____ 7. In desktop publishing, WYSIWYG stands for:
a. its founder   c. what you see is what you get
b. its energizer   d. its founder and energizer

TRUE/FALSE: Write the full word TRUE or FALSE in response to the following statements.

_____ 8. Styles of letters are called fonts.

_____ 9. The word photography means to write with ink.

_____ 10. Copying machines make use of photography and static electricity.

_____ 11. An orthographic drawing is one that has one or more vanishing points.

_____ 12. CAD stands for computer-aided design.

_____ 13. Johannes Gutenberg invented the printing press.

_____ 14. Computer-aided design is the fastest and most efficient way of drafting.

FILL IN THE BLANKS: In the blanks provided, write the correct answer to each question.

15. _____ _____ combine both typewriters and computer technologies.

16. _____ _____ lets a person turn out a book or newsletter page by page.

17. A _____ _____ _____ camera is like a twin lens reflex camera, except it has only one lens for both viewing and focusing.

IDENTIFICATION: Identify the following drawings as perspective, isometric or oblique:

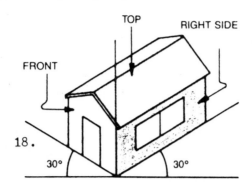

18. _____

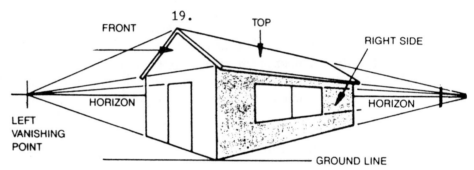

19. _____

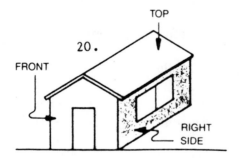

20. _____

GRAPHIC COMMUNICATION

ANSWER SHEET
///////// KEY WORDS \\\\\\\\\

1. isometric
2. CAD
3. orthographic
4. Screen printing
5. offset printing

6. daisy wheel
7. dot matrix
8. Relief printing
9. Gravure printing
10. perspective

///////// REVIEW \\\\\\\\\

FILL IN THE BLANKS:

1. a. Writing
   b. Technical drawing
   c. Printing
   d. Word processing
   e. Freehand drawing and sketching
   f. Photography
   or: Photocopying

2. a. The combination of design and drafting into one function saves time.
   b. The combination of other tasks with the CAD function reduces the chance of error.
   c. It cuts out most of the repetitive portions of hand drawing.
   d. New designs and changes to existing designs are much faster using CAD systems.
   e. Accuracy and consistency are improved from one drawing to the next by the use of a CAD system.

3. a. Light   b. Film   c. Camera   d. Chemicals   e. Dark area

4. a. View camera - has lens on one end and film and focusing on the other.
   b. Viewfinder - has separate viewfinder to compose picture.
   c. Twin lens reflex - one lens for viewing and one for focusing (light is reflected for focusing).
   d. Single lens reflex - same lens is used for both focusing and viewing.

5. a. Relief printing
   b. Screen printing
   c. Gravure printing
   d. Offset printing

6. a. Daisy wheel - has a daisy-shaped plastic print-head with a set of characters.
   b. Dot matrix - has a group of pins arranged in rectangular shapes which are pressed into a ribbon.
   c. Laser - makes use of a pulsing laser beam and a photosensitive drum.

TRUE/FALSE:

7. True
8. False
9. True
10. False
11. False
12. True

MULTIPLE CHOICE:

13. C
14. B
15. C
16. A
17. C

IDENTIFICATION:

18. Perspective
19. Oblique
20. Isometric
21. Orthographic

(Answers may vary and there may be additions to these suggested answers. Answers can depend upon the level of studies in the various disciplines for the group of students you are teaching.)

SCIENCE: In science you learn about light and color and how different colors are formed. This is important in both photography and printing newspapers, books, magazines, product labeling, etc.

MATH: As a photographer you will need to be able to work with numerical settings on the camera. You will also have to calculate mathematically some of the chemicals you will be using. You will also be working with other equipment that has numerical settings.

SOCIAL STUDIES: History has been recorded in the past by our ancestors who used dyes from nature to paint on the walls and ceilings of cliffs and caves. Others carved into stone symbols or figures recording their history. Still others made inks and pens and used these to write on leather, crude forms of paper, and fabrics.

LANGUAGE ARTS: A major part of being able to communicate is the ability to read, comprehend, and write or rewrite information.

ART: The more you know about art concepts and the greater the skill you develop in use of these, the more productive you are in the graphic communications field. If you were to own your own company, your sales would depend upon your knowledge and skill.

///////// *TEST* \\\\\\\\\

MULTIPLE CHOICE:

1. D
2. B
3. A
4. C
5. A
6. C
7. C

TRUE/FALSE:

8. True
9. False
10. True
11. False
12. True
13. True
14. True

FILL IN THE BLANKS:

15. Word processors
16. Desktop publishing
17. Single lens reflex

IDENTIFICATION:

18. Isometric
19. Perspective
20. Oblique

ELECTRONIC COMMUNICATION

Chapter 7, Text pages 166–206

KEY WORDS

Using the Key Words listed below, demonstrate your mastery of the Key Words by selecting the correct answer to complete each statement.

Amplifier	Electromagnetic wave	Tape head
ASCII	Fiber optic	Telecommute
Broadcast	Frequency	Teleconference
Cassette	Local area network	Telegraph
Compact disk	Modem	Telephone
Data communication	Network	Tracks
Distributed computing	Office automation	Uplink
Downlink	Stereo	Wavelength

1. The first electronic communication system to be widely used was the _____.

2. The most commonly used form of communication is the _____.

3. _____ is measured in cycles per second.

4. FM and AM radio, TV, microwave radio, and fiber optic communication all use _____ to send messages from the transmitter to the receiver.

5. A/An _____ occurs when a transmitting station on the earth sends signals to a satellite.

6. A/An _____ occurs when a satellite receives the signal and changes it to a different frequency and sends the signal back to earth.

7. A/An _____ _____ cable is made up of many thin strands of coated glass fibers that carry light for long distances without having to be amplified.

8. _____ _____ is communication between computers or between a computer and another peripheral device such as a printer or terminal.

9. The sending of larger, more difficult jobs from smaller computers to a larger one is called _____ _____.

10. _____ is a code that many computers and computer devices use.

ELECTRONIC COMMUNICATION

Chapter 7, Text pages 166–206

REVIEW

FILL IN THE BLANKS: In the blanks provided, write the correct answer to each question.

1. List seven types of transmitting and receiving systems.

   a. _____   e. _____

   b. _____   f. _____

   c. _____   g. _____

   d. _____

2. List four types of recording (ways of storing) systems.

   a. _____   c. _____

   b. _____   d. _____

3. What are four modern types of telecommunication services?

   a. _____   c. _____

   b. _____   d. _____

4. Systems that send messages immediately are called _____ and
   _____ systems.

5. Systems that store messages electronically are called _____
   _____.

MATCHING: Please match the statements on the right to the words at the left. Place the letter in the space provided.

____	6. Telegraph	A.	light-sensitive conductor device
____	7. Frequency	B.	transmission for one receiver
____	8. Long wavelengths	C.	measured in cycles per second
____	9. Short wavelengths	D.	radio signal with low frequency
____	10. Ionosphere	E.	sending pictures electronically
____	11. Point-to-point	F.	radio signal with high frequency
____	12. Network	G.	first electronic communication
____	13. Facsimile	H.	atmosphere's upper layer
____	14. Modem	I.	connect computers with telephone
____	15. Photodiode	J.	computers or computer devices joined together

NAME_____ SCHOOL_____

CLASS_____ PERIOD_____ DATE_____ SCORE_____

MULTIPLE CHOICE: Place the letter of the correct answer in the blank provided to the left of each question.

____16. In measuring frequency, one cycle per second is called a:
   a. photodiode
   b. modem
   c. Hertz
   d. signal

____17. Radio waves, microwaves, and light waves all occur at different frequencies and are called:
   a. tuners
   b. Hertz
   c. electromagnetic waves
   d. facsimiles

____18. Communication satellites are really radio relay stations, called:
   a. repeaters
   b. frequencies
   c. tuners
   d. electromagnetic

____19. A satellite that makes a complete revolution at the same time the earth does is said to have an orbit that is:
   a. stationary
   b. fixed
   c. revolutionary
   d. geosynchronous

____20. Communication between computers or from a computer to a terminal, printer, or other "peripheral" device is called:
   a. fiber optics
   b. data communication
   c. computer command
   d. microwave transmission

TRUE/FALSE: Write the full word TRUE or FALSE in response to the following statements.

_____ 21. Centralized computing is having some work done on small computers and difficult work sent to large computers at other locations.

_____ 22. ASCII is a common code that computers and computer devices use to communicate with each other.

_____ 23. Modulation is the turning of data signals into audio signals by use of a modem.

_____ 24. A laser disk has grooves that are read by laser needle, whereas a record has pits that are "read" by a fine steel needle.

_____ 25. Sound can be recorded on magnetic tape, phonograph records, and laser disks.

ELECTRONIC COMMUNICATION

Chapter 7, Text pages 166–206

INTERFACING WITH TECHNOLOGY

From the telegraph to telephone to fiber optics and satellite communication, the Information Age has made many changes in the ways and extent of our communications. We now have teleconferencing, electronic banking, and two-way cable TV. Along with the many benefits come responsibilities. To live in this Information Age, you will need to become technologically literate. Below you will be asked to give input regarding the need to have a well-rounded education.

SCIENCE: How will the study of science help us to understand the world of electronic communications?

_____

_____

MATH: How do you think a good background in math will help us live in this world so greatly impacted by electronic communications?

_____

_____

SOCIAL STUDIES: What effect have all the advances in electronic communications had on social studies?

_____

_____

LANGUAGE ARTS: What is the difference between being literate and being technologically literate?

_____

_____

ART: How will a knowledge of basic art concepts help you make decisions about purchases of electronic communication systems such as telephones, radio, and television?

_____

_____

ELECTRONIC COMMUNICATION

Chapter 7, Text pages 166–206

TEST

TRUE/FALSE: Write the full word TRUE or FALSE in response to the following statements.

_____ 1. Facsimile (FAX) is a method of sending pictures electronically.

_____ 2. A modem is a device equipped with grooves and pits which are read by a laser.

_____ 3. A number of computers joined together is called a network.

_____ 4. Modern telecommunication systems include teleconferencing, telecommuting, electronic banking, and two-way cable TV.

_____ 5. Frequency is measured in cycles per minute.

MULTIPLE CHOICE: Place the letter of the correct answer in the blank provided for each question.

_____ 6. A satellite that makes a complete revolution at the same time the earth does is said to have an orbit that is:
    a. stationary   c. revolutionary
    b. fixed   d. geosynchronous

_____ 7. Radio waves, microwaves, and light waves all occur at different frequencies and are called:
    a. tuners   c. electromagnetic waves
    b. Hertz   d. facsimiles

_____ 8. Communication between computers or from a computer to a terminal, printer, or other "peripheral" device is called:
    a. fiber optics   c. computer command
    b. data communication   d. microwave transmission

_____ 9. Sound can be recorded on magnetic tape, phonograph records and:
    a. laser disks   c. compact disks
    b. amplifiers   d. both "a" and "c"

_____ 10. Radio signals with high frequencies are said to have wavelengths that are:
    a. long   c. short
    b. electrical   d. magnetic

ELECTRONIC COMMUNICATION

Chapter 7, Text pages 166–206

ANSWER SHEET

/////////// KEY WORDS \\\\\\\\\\\\

1. telegraph
2. telephone
3. Frequency
4. electromagnetic waves
5. uplink
6. downlink
7. fiber optic
8. Data communications
9. distributed computing
10. ASCII

/////////// REVIEW \\\\\\\\\\\\

FILL IN THE BLANKS:

1. a. Telegraph
   b. Telephone
   c. Radio
   d. Television
   e. Microwave
   f. Satellite
   g. Fiber optic

2. a. Magnetic recording tape
   b. Magnetic recording disk
   c. Phonograph record
   d. Optical disk

3. a. Teleconferencing
   b. Telecommuting
   c. Electronic banking
   d. Two-way Cable TV

4. transmitting and receiving

5. recording systems

MATCHING:

6. G
7. C
8. D
9. F
10. H
11. B
12. J
13. E
14. I
15. A

MULTIPLE CHOICE:

16. C
17. C
18. A
19. D
20. B

TRUE/FALSE:

21. False
22. True
23. True
24. False
25. True

/////////// INTERFACING WITH TECHNOLOGY \\\\\\\\\\\\

(Answers may vary and there may be additions to these suggested answers. Answers can depend upon the level of studies in the various disciplines for the group of students you are teaching.)

SCIENCE: Science gives us a background in electricity and electronics which is the basis for electronic communications.

MATH: Even though there are many electronic devices surrounding us, we must still be able to calculate change, determine if we can afford products, and how much we can afford. Many simple day-to-day calculations require being able to use math. There are times we must know how to use measurements so that we get what we ask and pay for.

SOCIAL STUDIES: The history of communications has greatly changed in the last years. The number of inventions to be recorded in history has grown exponentially. History now shows technologies such as satellite, microwave, and teleconferencing, which bring society closer together.

LANGUAGE ARTS: To be literate is to be able to read and write; to be technologically literate means to be able to read and understand and have a working knowledge of the technologies.

ART: When purchasing communication equipment, we need to be able to choose among a wide variety of colors, styles, and shapes so that it coordinates with the area in which the equipment will be placed.

///////// **TEST** \\\\\\\\\\

TRUE/FALSE:

1. True
2. False
3. True
4. True
5. False

MULTIPLE CHOICE:

6. D
7. C
8. B
9. D
10. C

PROCESSING MATERIALS

Chapter 8, Text pages 208–245

KEY WORDS

Using the Key Words listed below, demonstrate your mastery of the Key Words by selecting the correct answer to complete each statement.

Brittle	Ductile	Industrial materials	Separating
Casting	Elastic	Insulator	Shaping
Ceramics	Extruding	Plasticity	Shearing
Coating	Fastening	Polymers	Tension
Combining	Ferrous metals	Pressing	Thermal
Composites	Forging	Properties of	Thermoplastics
Compression	Forming	materials	Thermoset plastics
Conditioning	Gluing	Raw materials	Torsion
Conductor	Grinding	Recycle	Toughness
Drilling	Heat-treating	Sawing	Turning

1. Cotton fiber and wood are known as _____ _____.

2. Metals with a content of more than 50% iron are called _____ _____.

3. Long chains of molecules used to make plastic are called _____.

4. PVC plumbing pipe, plastic bags, and nylon that softens when heated are known as _____ plastics.

5. Plastic cups and dishes, bakelite, and Formica do not soften when heated and are known as _____ plastics.

6. The process of changing a material by shearing, sawing, drilling, and shaping is known as _____.

7. The process of changing a material without cutting it is known as _____.

8. The process of changing a material by fastening, gluing, and making composites is known as _____.

9. The process of changing a material by firing, heat treating, and magnetizing is known as _____.

10. The process of squeezing softened material through an opening is known as _____.

11. The process of heating and applying large amounts of force on a material to form its shape is known as _____.

12. Ripping and crosscutting are separation processes known as _____.

13. _____ is a separating process which is done with the use of hard materials known as abrasives that are crushed and adhered to a surface.

14. The combining process of _____ is done using nails, screws, and rivets.

15. The combining process known as _____ produces surfaces that are galvanized, electroplated, or anodized.

16. A/An _____ material such as plywood and fiberglass is made by combining several materials.

17. The _____ _____ _____ include strength, hardness, appearance, ability to conduct electricity, and resistance to corrosion.

18. _____ is the force that pushes on or squeezes a material.

19. _____ is a force that pulls on a piece of material, whereas torsion is the twisting of the material.

20. We need to _____ products of materials that will not decompose, like plastic and glass.

PROCESSING MATERIALS

Chapter 8, Text pages 208–245

REVIEW

FILL IN THE BLANKS: In the blanks provided, write the correct answer to each question.

1. List and give examples of four ways of processing materials.

    a. _____   _____

    b. _____   _____

    c. _____   _____

    d. _____   _____

2. List and briefly describe five ways materials can be formed.

    a. _____   _____

    b. _____   _____

    c. _____   _____

    d. _____   _____

    e. _____   _____

3. List six separating processes.

    a. _____   d. _____

    b. _____   e. _____

    c. _____   f. _____

4. List the five combining processes of materials and describe each briefly.

    a. _____   _____

    _____

    b. _____   _____

    _____

    c. _____   _____

    _____

    d. _____   _____

    _____

    e. _____   _____

    _____

5. List four types of conditioning processes.

    a. _____   c. _____

    b. _____   d. _____

6. List the five characteristics or properties of materials.

    a. _____   d. _____

    b. _____   e. _____

    c. _____

MATCHING: Please match the statements on the right to the words at the left. Place the letter in the space provided.

____ 7. Ferrous
____ 8. Alloy
____ 9. Hardwood
____10. Ceramic
____11. Polymers
____12. Extruding
____13. Composite
____14. Thermoset
____15. Thermal
____16. Toughness

A. plastics with long chains of molecules
B. process where softened material is squeezed through an opening
C. material produced by combining several materials together
D. ability of a material to absorb energy without breaking
E. metals made of more than 50% iron
F. property of a material referring to heat
G. combinations of metals
H. wood from trees with broad leaves
I. made from inorganic materials (clay)
J. plastic that will not soften when heated (bakelite and Formica)

TRUE/FALSE: Write the full word TRUE or FALSE in response to the following statements.

_____ 17. Primary industries produce raw materials.

_____ 18. Plastic that is heated along with sulfur dust is called vulcanized.

_____ 19. Aluminum, tin, and iron are non-ferrous metals.

_____ 20. Glass is a ceramic material.

_____ 21. Thermoplastics are plastics that soften when heated and retain original hardness when cooled.

_____ 22. Separating processes include blow molding and vacuum forming.

MULTIPLE CHOICE: Place the letter of the correct answer in the blank provided to the left of each question.

____23. A forming process where metal is heated and hammered into shape:
    a.   casting   c.   forging
    b.   extruding   d.   blow molding

____24. Process where thin sheets of plastic are heated and air blows the plastic into a mold:
    a.   casting   c.   forging
    b.   extruding   d.   blow molding

____25. A separating process where knife-like blades are used:
    a.   shearing   c.   sawing
    b.   grinding   d.   drilling

____26. A separating process that makes use of crushed hard materials called abrasives:
    a.   shearing   c.   sawing
    b.   grinding   d.   drilling

____27. Materials can be fastened together by:
    a.   screws and nails   c.   heat
    b.   shearing   d.   Both "a" and "c"

____28. Materials that beautify or protect the surface are called:
    a.   coatings   c.   gluing
    b.   fasteners   d.   riveters

____29. The coating process that coats steel with zinc to keep the steel from rusting is:
    a.   both "d" and "c"   c.   electroplating
    b.   anodizing   d.   galvanizing

____30. A conditioning process where steel is heated to a cherry-red color and quickly cooled in water and then heated to an intermediate temperature and quickly cooled is called:
    a.   annealing   c.   hardening
    b.   tempering   d.   magnetizing

____31. Wood is classified as hardwood, softwood, and:
    a.   turned wood   c.   manufactured wood
    b.   knotty   d.   paste

____32. The property that refers to a material's ability to reflect light:
    a. thermal   c.   electrical
    b. optical   d.   magnetic

PROCESSING MATERIALS

Chapter 8, Text pages 208–245

INTERFACING WITH TECHNOLOGY

Processing materials makes them more useful and valuable. As we process materials to make the products that supply our needs and wants, we are faced with depletion of resources, waste materials, and pollution.   Below you will be asked to give input considering these thoughts.

SCIENCE: How will the study of science help us in our need for resources required in the processing of materials?

_____

_____

MATH: We use wood, fossil fuels, and metal ores. How will a background in math help you make decisions regarding the regulation of methods of using these resources?

_____

_____

SOCIAL STUDIES: Compare the use of resources of the Information Age to that of the Industrial Age.

_____

_____

LANGUAGE ARTS: After reading and studying the materials in this chapter, why is it so important that you can read?

_____

_____

ART: How is art used in the processing of materials?

_____

_____

PROCESSING MATERIALS
Chapter 8, Text pages 208–245

TEST

TRUE/FALSE: Write the full word TRUE or FALSE in response to the following statements.

_____ 1. Ferrous metals contain more than 50% iron.

_____ 2. Separating processes include welding, gluing, and soldering.

_____ 3. Conditioning processes include heat treating, chemical conditioning, and firing.

_____ 4. Using nails, screws, and brads are mechanical means of fastening materials.

_____ 5. Blow molding and vacuum forming are separating processes.

_____ 6. Thermoset plastics do not soften when heated.

_____ 7. Extruding is heating and hammering a metal part into shape.

_____ 8. Particle board is known as manufactured board.

MATCHING: Please match the statements on the right to the words at the left. Place the letter in the space provided.

____ 9. Galvanize   A. ability of a material to absorb energy without breaking
____10. Composite   B. plastics with long chains of molecules
____11. Polymers   C. process of cutting with blades
____12. Toughness   D. a form of ceramics
____13. Glass   E. wood from trees with broad leaves
____14. Shearing   F. property of material referring to heat
____15. Hardwood   G. coating steel with zinc to stop rust
____16. Thermal   H. material made by combining several materials

MULTIPLE CHOICE: Place the letter of the correct answer in the blank provided to the left of each question.

____17. Firing and heat treating are examples of:
   a. forming processes   c. separating processes
   b. combining processes   d. conditioning processes

____18. Casting, pressing, and extruding are examples of:
   a. forming processes   c. separating processes
   b. combining processes   d. conditioning processes

____19. Shearing, sawing, and grinding are examples of:
   a. forming processes   c. separating processes
   b. combining processes   d. conditioning processes

____20. Coating, gluing, and fastening are examples of:
   a. forming processes   c. separating processes
   b. combining processes   d. conditioning processes

____21. Soldering and welding are examples of:
    a.   forming processes   c.   separating processes
    b.   combining processes   d.   conditioning processes

____22. The property that refers to a material's ability to reflect light is:
    a.   optical   c.   thermal
    b.   magnetic   d.   electrical

____23. Blow molding and vacuum forming are examples of:
    a.   separating processes   c.   conditioning processes
    b.   forming processes   d.   combining processes

____24. Shaping and turning are examples of:
    a.   separating processes   c.   conditioning processes
    b.   forming processes   d.   combining processes

____25. A conditioning process where steel is heated to a cherry-red color and quickly cooled in water and then heated to an intermediate temperature and quickly cooled is called:
    a.   annealing   c.   tempering
    b.   magnetizing   d.   hardening

PROCESSING MATERIALS

Chapter 8, Text pages 208–245

ANSWER SHEET
////////// KEY WORDS \\\\\\\\\\

1. raw materials
2. ferrous metals
3. polymers
4. thermoplastic
5. thermoset
6. separating
7. forming
8. combining
9. conditioning
10. extrusion
11. forging
12. sawing
13. Grinding
14. fastening
15. coating
16. composite
17. properties of materials
18. Compression
19. Tension
20. recycle

////////// REVIEW \\\\\\\\\\

FILL IN THE BLANKS:

1. a. Forming. . .casting, pressing, forging, extruding, and blow molding and vacuum forming.
    b. Separating. . .shearing, sawing, drilling, grinding, shaping, and turning.
    c. Combining. . .mechanical fastening, gluing, fastening with heat, coating, and making composites.
    d. Conditioning. . .heat-treating, mechanical conditioning, chemical conditioning, and firing.

2. a. Casting. . .pouring material into a mold.
    b. Pressing. . .pouring exact amount of material into a mold and pressing into shape.
    c. Forging. . .heating and hammering metal into shape.
    d. Extruding. . softening a material and squeezing it through an opening to form a shape.
    e. Blow molding. . .thin plastic sheet is heated until softened and either blown or vacuum-formed into shape.

3. a. Shearing   c. Shaping   e. Grinding
    b. Sawing   d. Drilling   f. Turning

4. a. Mechanical. . .using nails, screws, and rivets to fasten two materials together.
    b. Heat. . .using heat to solder or weld materials.
    c. Gluing. . .using glues to create a chemical bond between itself and the materials being glued (epoxy, SuperGlue, white glue).
    d. Coating. . .using paint, glazes, waxes, or processes such as electroplating, anodizing, or galvanizing.
    e. Composites. . .combining several materials to form a new material with special properties.

5. a. Heat-treating   c. Chemical conditioning
    b. Firing   d. Mechanical conditioning

6. a. Mechanical   d. Thermal
    b. Electrical   e. Optical
    c. Magnetic

MATCHING:	**TRUE/FALSE:**	**MULTIPLE CHOICE:**
7. E	17. True	23. C
8. G	18. False	24. D
9. H	19. False	25. A
10. I	20. True	26. B
11. A	21. True	27. A
12. B	22. False	28. A
13. C		29. D
14. J		30. B
15. F		31. C
16. D		32. B

///////// INTERFACING WITH TECHNOLOGY \\\\\\\\\\

(Answers may vary and there may be additions to these suggested answers. Answers can depend upon the level of studies in the various disciplines for the group of students you are teaching.)

SCIENCE: With a good background in the sciences, we will be able to make wise choices in the use of resources. We will know that resources are renewable and nonrenewable and why. This will develop in us a responsible attitude toward the use and recycling of resources.

MATH: Math helps us to develop an awareness of the large use of our resources in the processing of materials. Knowing how much is used and how much waste and pollution results enables us to make responsible decisions regarding the regulation of use of resources.

SOCIAL STUDIES: During the Industrial Age, we began to mass produce products and use more resources; however, in the Information Age, we have an exponential increase in production. We are using resources very quickly; however, there are greater efforts to curb pollution and attempt to recycle materials. During the Industrial Age, recycling and pollution were not considered problems.

LANGUAGE ARTS: If you cannot read, you will not be able to read or comprehend labels and instruction manuals. An awareness of the products you use enables you to take charge of the product and maintain or choose wisely someone to repair or maintain that product.

ART: The elements of art are always a factor in the design of materials to be processed into usable products.

///////// T E S T \\\\\\\\\\

TRUE/FALSE:	**MATCHING:**	**MULTIPLE CHOICE:**
1. True	9. G	17. D
2. False	10. H	18. A
3. True	11. B	19. C
4. True	12. A	20. B
5. False	13. D	21. B
6. True	14. C	22. A
7. False	15. E	23. B
8. True	16. F	24. A
		25. C

PRODUCTION SYSTEMS: MANUFACTURING

Chapter 9, Text pages 246–275

KEY WORDS

Using the Key Words listed below, demonstrate your mastery of the Key Words by selecting the correct answer to complete each statement.

Assembly line	Flexible manufacturing	Numerical control
Automation	Interchangeable	Production
CAD/CAM	Inventor	Prototype
CIM	Just-In-Time	Quality control
Custom-made	manufacturing	Robot
Entrepreneur	Manufacturing	Uniformity
Factory	Mass production	Union
Feedback control		

1. _____ is making goods in a workshop or factory.

2. The production of goods in large quantities by groups of workers in factories is called _____ _____.

3. A person who comes up with a good idea and makes money with that idea is called a/an _____.

4. The process of controlling machines automatically is called _____.

5. _____ combines the manufacturing, design, and business functions of a company under the control of a computer system.

6. _____ _____ is the production of small amounts of products where the same production line can be used for all versions of a product.

7. Parts and materials arrive at the factory when needed in _____ _____.

8. _____ are better than ordinary machines, because they can be reprogrammed to perform different tasks.

9. The _____ _____ is a system in which the item is moved quickly from one work station to the next.

10. _____ technologies fill people's needs and wants by manufacturing and construction systems.

PRODUCTION SYSTEMS: MANUFACTURING

Chapter 9, Text pages 246–275

REVIEW

FILL IN THE BLANKS: In the blanks provided, write the correct answer to each question.

1. A_____ holds and guides an item being processed. A_____ is used to keep the item being processed in the proper position.

2. _____ were people who specialized in manufacture of one item (candlemakers, silversmiths, and tailors).

3. The_____ _____ grew out of the inventions during the Industrial Revolution.

4. In 1798, _____ _____ started one of the earliest mass-production assembly lines.

5. _____ parts are those that are mass-produced to be exactly alike, making them interchangeable.

6. An_____ is a person who comes up with a good idea and uses that idea to make money.

7. An _____ is a person who comes up with a totally new idea.

8. An _____ is a person who improves an invention and leads to other uses which can start new industries or change existing ones.

9. The seven resources for manufacturing systems are:

    a. _____   e. _____

    b. _____   f. _____

    c. _____   g. _____

    d. _____

10. _____ _____ is producing products in small quantities for customers with special needs.

11. _____ _____ is making sure finished products are usable and quality remains high during manufacture.

MULTIPLE CHOICE: Place the letter of the correct answer in the blank provided to the left of each question.

_____12. Before a company goes into production it conducts:
     a.   quality circles   c.   market research
     b.   prototyping   d.   research and development

_____13. In manufacturing, research and development (R & D) is conducted by:
     a.   quality circles   c.   product designers
     b.   engineers   d.   both "b" and "c"

_____14. The step following the agreement by engineers and businessmen on the design of a
product is that of:
    a.   designing   c.   flow-charting
    b.   testing   d.   prototyping

_____15. Once a product is tested, the production line is set up by diagramming the sequence
of steps through the use of a:
    a.   flow-chart   c.   prototype
    b.   test   d.   design

_____16. The process of controlling machines automatically is called:
    a.   automation   c.   prototyping
    b.   modeling   d.   simulation

COMPLETE THE CHART: The chart below relates to terminology in manufacturing. Fill in
the missing information.

ABBREVIATION	MANUFACTURING TERMINOLOGY	DEFINITION
17. CAD		
18. CAD/CAM		
19. CIM		
20. JIT		

TRUE/FALSE: Write the full word TRUE or FALSE in response to the following statements.

_____ 21. Robots are advanced automated machines.

_____ 22. Productivity is producing less at lower costs.

_____ 23. Observing workers closely while they are working is called scientific
management.

_____ 24. Money needed to buy land, equipment, and buildings, and to pay workers,
is called cogeneration.

_____ 25. Running tests on a computer screen is called simulation.

PRODUCTION SYSTEMS: MANUFACTURING

Chapter 9, Text pages 246–275

INTERFACING WITH TECHNOLOGY

The two production systems are manufacturing and construction. Construction is building a structure on site, whereas manufacturing is making goods in a workshop or factory. Automation has greatly increased productivity. It has also produced many changes. Why do you think we need to have a good background in the below subjects?

SCIENCE: How does the study of science relate to the manufacture of goods? How will this affect us?

_____

_____

MATH: How do you think math is used in the manufacture of products?

_____

_____

SOCIAL STUDIES: What major event in history changed the manufacturing of products? What type of change came about?

_____

_____

LANGUAGE ARTS: If you are a consumer of the many products that are manufactured, why is it so important that you be able to read and comprehend that which you read?

_____

_____

ART: If you studied art and truly enjoyed it and wanted to use it in industry, what type jobs do you think require artistic abilities?

_____

_____

PRODUCTION SYSTEMS: MANUFACTURING

Chapter 9, Text pages 246–275

TEST

MULTIPLE CHOICE: Place the letter of the correct answer in the blank provided to the left of the question.

_____ 1. One of the earliest inventions was the:
- a. telegraph
- c. firing of clay
- b. water wheel
- d. steam engine

_____ 2. Individuals who specialized in and made one product at a time:
- a. inventors
- c. mass producers
- b. craftspeople
- d. R & D

_____ 3. Making sure that finished products are usable and quality remains high during manufacture is called:
- a. quality circles
- c. quality control
- b. simulation
- d. both "b" and "c"

_____ 4. One who improves on an invention is called an:
- a. entrepreneur
- c. inventor
- b. innovator
- d. conductor

_____ 5. Items that are manufactured and ready to be used as needed is called:
- a. JIT
- c. CAD
- b. CAM
- d. JAT

TRUE/FALSE: Write the full word TRUE or FALSE in response to the following statements.

_____ 6. The process of controlling machines automatically is called automation.

_____ 7. Robots are advanced automated machines.

_____ 8. Scientific management is allowing scientists to perform the jobs designed for robots.

_____ 9. Manufacturing makes use of five types of technological systems.

_____ 10. CAD makes use of a modem, computer, and a robot.

PRODUCTION SYSTEMS: MANUFACTURING

Chapter 9, Text pages 246–275

ANSWER SHEET

/////////// KEY WORDS \\\\\\\\\\\

1. Manufacturing
2. mass production
3. entrepreneur
4. automation
5. CIM

6. Flexible manufacturing
7. Just-In-Time manufacturing
8. Robots
9. assembly line
10. Production

/////////// REVIEW \\\\\\\\\\\

FILL IN THE BLANKS:

1. jig, fixture
2. Craftspeople
3. factory system
4. Eli Whitney
5. Standardized
6. entrepreneur
7. inventor
8. innovator
9. a. People   e. Energy
   b. Information   f. Capital
   c. Materials   g. Time
   d. Tools and machines
10. Custom made
11. Quality control

MULTIPLE CHOICE:

12. C
13. D
14. D
15. A
16. A

COMPLETE THE CHART:

17.	CAD	Computer-aided design	Using computers to draw designs.
18.	CAD/CAM	Computer-aided design and Computer-aided manufacturing	Design and design input are fed into a computer that communicates to a machine tool for production.
19.	CIM	Computer-integrated manufacturing	System that designs, stores data, controls machines, schedules purchase and delivery, and does billing and accounting.
20.	JIT	Just-In-Time manufacturing	Raw materials and parts arrive at the place and time needed.

TRUE/FALSE

21. True
22. False
23. False
24. False
25. True

(Answers may vary and there may be additions to these suggested answers. Answers can depend upon the level of studies in the various disciplines for the group of students you are teaching.)

SCIENCE: Many of the goods that are produced are made of synthetics that are made by mixing materials to form a new substance. In the science field, we study resources such as chemicals, chemical compositions, and synthetics. Since some of our natural resources are limited, we need to find ways to replace these. Research scientists work constantly to develop replacements and make use of recycled materials. The products we use depend on scientific research.

MATH: Market research is based on the numbers and kinds of responses given by consumers. These data are run through computers for studies. The number of products is controlled by research. Also, many products are assigned serial and model numbers which can be coded.

SOCIAL STUDIES: The Industrial Revolution with its factory systems changed the manner in which products were produced. We changed from a society of craftspeople to factory workers.

LANGUAGE ARTS: Products have labels which contain important information concerning correct use and safety. Instruction manuals must be read for purposes of maintaining and doing simple repairs to products. If we are able to read and comprehend what we read, we can make wiser decisions in the purchase of products, and we can read warranties and contracts.

ART: All products are the result of someone who has studied and is experienced in designing products. You can enter the field of advertisement of products, become a design engineer, or work as a designer.

////////// **TEST** \\\\\\\\\\

MULTIPLE CHOICE:

1. C
2. B
3. C
4. B
5. A

TRUE/FALSE:

6. True
7. True
8. False
9. False
10. False

PRODUCTION SYSTEMS: CONSTRUCTION

Chapter 10, Text pages 276-301

KEY WORDS

Using the Key Words listed below, demonstrate your mastery of the Key Words by selecting the correct answer to complete each statement.

Arch	Engineer	Macadam	Structure
Architect	Estimator	Mortgage	Superstructure
Cement	Forms	Prefabricate	Surveyor
Concrete	Foundation	Specifications	
Construction	General contractor	Steel	

1. Civil _____ work with architects by preparing drawings and plans for the building framework and foundation with sizes and strengths of materials to be used.

2. The _____ figures the cost of the project and submits a bid.

3. _____ _____ own their own companies and hire workers and oversee part or all of a project.

4. The _____ is the part of the structure that supports the weight of the structure.

5. The _____ is the part of the structure that is above the ground that is usable.

6. Parts of a structure are _____ in a factory and moved to the construction site.

7. A loan for a building is called the _____.

8. _____ is used for the framework of skyscrapers, bridges, and towers.

9. The _____ are the data like the materials to be used, the way the foundation will be built, and even the kinds of trees and bushes to be planted around the structure.

10. The _____ designs the building, showing how a structure will be built and where it will be placed on the site.

PRODUCTION SYSTEMS: CONSTRUCTION

Chapter 10, Text pages 276-301

REVIEW

MULTIPLE CHOICE: Place the letter of the correct answer in the blank provided to the left of each question.

_____ 1. In the construction industry, the designing and planning is done by the:
     a.   land owner   c.   architect and engineers
     b.   estimator   d.   civil engineer

_____ 2. Drawings and plans along with the size and strength of construction materials and details about utilities are completed by the:
     a.   civil engineers   c.   project managers
     b.   estimators   d.   general contractors

_____ 3. Individuals who own their own construction company and hire and work directly with the owner, architects, engineers, and craftspeople are known as:
     a.   civil engineers   c.   project managers
     b.   estimators   d.   general contractors

_____ 4. Cost proposals called bids are prepared by:
     a.   civil engineers   c.   project managers
     b.   estimators   d.   general contractors

_____ 5. Plans for bridges, roads, dams, tunnels, and towers are developed by:
     a.   civil engineers   c.   project engineers
     b.   estimators   d.   general contractors

MATCHING: At the right are seven resources needed in construction. Match these to the definitions to the left.

_____ 6. Cranes, bulldozers, and tools   A. People
_____ 7. Contractors and architects   B. Time
_____ 8. Loans or governmental aid   C. Materials
_____ 9. Used to produce materials   D. Tools and machines
____10. Concrete, wood, brick, and glass   E. Capital
____11. Plans and specifications   F. Energy
____12. Periods of months up to years   G. Information

FILL IN THE BLANKS: In the spaces provided, write the correct answers to each question.

13. _____ refers to producing a structure on a site.

14. The construction process involves the following steps:

a. _____

b. _____

c. _____

d. _____

e. _____

15. Seven types of structures include:

a. _____   e. _____

b. _____   f. _____

c. _____   g. _____

d. _____

TRUE/FALSE: Write the full word TRUE or FALSE in response to the following statements.

_____16. Parts of a building constructed in a factory are said to be prefabricated.

_____17. A structure includes a foundation and a superstructure.

_____18. The cantilever bridge is used to bridge wide spans.

_____19. Building construction includes residential, commercial, dimensional, and industrial.

_____20. A building foundation supports the weight of a structure.

PRODUCTION SYSTEMS: CONSTRUCTION

Chapter 10, Text pages 276–301

INTERFACING WITH TECHNOLOGY

Construction is the building of a structure on site. Structures include bridges, buildings, dams, harbors, roads, towers, and tunnels. Every day we make use of structures. As you become adults, you will be making decisions about structures, since you will be the one who will be using and paying for these structures. To prepare yourself to make wise choices, you will need to become well informed. How can you prepare yourself?

SCIENCE: What do you learn in science that relates to structures?

_____

_____

MATH: How do you think math will help you personally?

_____

_____

SOCIAL STUDIES: Where did early people live? What were some of the earliest structures like?

_____

_____

LANGUAGE ARTS: If you were going to buy a house or lease an apartment, condominium, or house, why would you need to be able to read and comprehend what you read?

_____

_____

ART: If you studied art and truly enjoyed it and wanted to use it in industry, what type of jobs do you think would require your artistic abilities?

_____

_____

NAME_____ SCHOOL_____

CLASS_____ PERIOD_____ DATE_____ SCORE_____

PRODUCTION SYSTEMS: CONSTRUCTION

Chapter 10, Text pages 276–301

TEST

TRUE/FALSE: Write the full word TRUE or FALSE in response to the following statements.

_____ 1. Construction systems process the seven technological resources to provide a structure.

_____ 2. Financing for a project, whether through loans or government aid, is called prefabrication.

_____ 3. The unusable part of a structure is the superstructure.

_____ 4. The technological resource of information includes the plans and specifications.

_____ 5. Civil engineers develop plans for dams, bridges, roads, tunnels, towers, and airports.

MULTIPLE CHOICE: Place the letter of the correct answer in the blank provided for each question.

____ 6. The first step in the construction process is:
 a. preparing the site
 b. selecting the site
 c. paying the loan
 d. building the foundation

____ 7. The bridge used to bridge wide spans is the:
 a. beam bridge
 b. arch bridge
 c. suspension bridge
 d. cantilever bridge

____ 8. The four categories of buildings are the residential, commercial, industrial, and the:
 a. individual home
 b. institutional
 c. factories
 d. hotels and housing

____ 9. The purpose of the building foundation is to:
 a. encase the windows
 b. insulate walls
 c. provide bids
 d. support the structure

____ 10. Wood, glass, steel, brick, and concrete are called:
 a. materials
 b. prefabrications
 c. information systems
 d. superstructures

PRODUCTION SYSTEMS: CONSTRUCTION

Chapter 10, Text pages 276-301

ANSWER SHEET

///////// KEY WORDS \\\\\\\\\

1. engineers
2. estimator
3. General contractors
4. foundation
5. superstructure
6. prefabricated
7. mortgage
8. Steel
9. specifications
10. architect

///////// REVIEW \\\\\\\\\

MULTIPLE CHOICE:

1.	C	4.	B
2.	A	5.	A
3.	D		

MATCHING:

6.	D	10.	C
7.	A	11.	G
8.	E	12.	B
9.	F		

FILL IN THE BLANKS:

13. Construction

14. a. Choosing and preparing the site
    b. Building the foundation
    c. Building the superstructure
    d. Installing utilities
    e. Finishing and enclosing the outside surfaces

15. a. Bridges
    b. Buildings
    c. Dams
    d. Harbors
    e. Roads
    f. Towers
    g. Tunnels

TRUE/FALSE:

16. True   17. True   18. False   19. False   20. True

///////// INTERFACING WITH TECHNOLOGY\\\\\\\\\

(Answers may vary and there may be additions to these suggested answers. Answers can depend upon the level of studies in the various disciplines for the group of students you are teaching.)

SCIENCE: In science we study about the elements of nature, such as wind, rain, and temperatures, and their effects. Nature greatly impacts the way structures are built.

MATH: When the time comes to lease or purchase a place to live, you will need to know how to balance a budget. You will need to be able to comparison shop for appliances for the place in which you will live. If repairs need to be made, your ability to estimate the cost will enable you to make wise decisions in choices when you either do it yourself or ask someone else to do the repairs.

SOCIAL STUDIES: The earliest people lived in caves or in the brush. The first buildings were simple structures such as tepees made from animal hides stretched over a wooden frame.

LANGUAGE ARTS: When you lease or buy a place to live, you will need to be able to read and comprehend contracts. Once you commit yourself by signing a contract, you have to follow the terms of the contract, whether you understood it or not.

ART: Artistic abilities are needed when you become an architect and even a civil engineer. As a home owner or one who leases a dwelling, artistic skills will help you to tastefully decorate your home.

////////// TEST \\\\\\\\\\

TRUE/FALSE:

1. True
2. False
3. False
4. True
5. True

MULTIPLE CHOICE:

6. B
7. C
8. B
9. D
10. A

BUILDING A STRUCTURE
Chapter 11, Text pages 302–341

KEY WORDS

Using the Key Words listed below, demonstrate your mastery of the Key Words by selecting the correct answer to complete each statement.

Asphalt	Frost line	National Electrical	Stud
Bottom plate	Girder	Code	Slab foundation
Building permit	Header	Pitch	Subfloor
Circuit breaker	Insulation	Plumb	Top plate
Floor joists	Manufactured	Plumbing	Truss
Footing	housing	Rafter	Vapor barrier
Foundation wall	Mason	Sheathing	Wind load
Framing	Mortar	Sill plate	

1. A_____ _____ allows the builder to begin work.

2. The base of the foundation wall made of concrete is called the _____.

3. Concrete block foundations are laid by a _____.

4. _____ is a mixture of cement, lime, sand, and water which acts like a glue, bonding the concrete blocks to the footing and to each other.

5. The depth to which the ground freezes in the winter is called the _____ _____.

6. The strong piece of wood nailed across the top of the window or door opening when windows are framed is called a _____.

7. The framed wall is covered with wall _____ that is made from plywood, particle board, wooden planks, or rigid foam board.

8. The _____ is how steep the slope of a roof must be.

9. The material that does not conduct heat very well that is placed between the walls is called _____.

10. The waterproof material placed over insulation acts as a _____ _____ to prevent condensation.

BUILDING A STRUCTURE
Chapter 11, Text pg. 302–341

REVIEW

FILL IN THE BLANKS: In the blanks provided, write the correct answer to each corrections.

1. Building construction involves the following three major steps:

    a. _____

    b. _____

    c. _____

2. The preconstruction phase involves these two steps:

    a. _____

    b. _____

3. The two parts of house foundations include:

    a. _____

    b. _____

4. _____ _____ are buildings that are used for banks, stores, and offices.

5. The _____ _____ _____ establishes safe methods of installing electrical wiring and equipment.

MULTIPLE CHOICE: Place the letter of the correct answer in the blank provided to the left of each question.

____ 6. Building permits are issued to builders only when plans comply with the local:
   a. union
   b. building code
   c. city councils
   d. drainage ditches

____ 7. The foundation wall is built upon the:
   a. footing
   b. studs
   c. drain system
   d. utilities

____ 8. Walls designed to support weight from above are:
   a. dry walls
   b. insulating walls
   c. sheathing
   d. load bearing walls

____ 9. Wall sheathing is used to make the house frame:
   a. more rigid
   b. weather resistant
   c. more attractive
   d. smaller

____ 10. Roofing materials are applied directly on top of:
   a. floor joists
   b. wall studs
   c. decking
   d. ceiling joists

____ 11. The plumbing, heating, and electrical systems are called:
    a. decking   c. utilities
    b. circuitry   d. joists

____ 12. Asphalt shingles, wood shingles, and shakes are called:
    a. roofing materials   c. utilities
    b. foundation materials   d. sheathing materials

____ 13. The extra wire in cable used in homes is called the:
    a. hot wire   c. ground wire
    b. alternate wire   d. parallel circuit wire

____ 14. Interior finishing is often referred to as:
    a. interior decorating   c. finishing
    b. drywall construction   d. plumbing

____ 15. Housing can be done on site or it can be:
    a. built stick-by-stick   c. manufactured
    b. produced in factories   d. both "b" and "c"

TRUE/FALSE: Write the full word TRUE or FALSE in response to the following statements.

_____ 16. The purpose of the roof is to protect the house against weather, prevent heat from escaping, and to beautify it.

_____ 17. Manufactured houses are built on site.

_____ 18. Wind does not influence the design and construction of skyscrapers.

_____ 19. When 2 X 4 studs are used, they are spaced 16 inches on center.

_____ 20. The Truss-frame system (TFS) makes use of a roof truss, two wall studs, and a floor truss assembled in three units.

_____ 21. Exterior finishing refers to finishing the inside walls and ceilings.

_____ 22. The hot and cold water supply and the drainage is known as the domestic plumbing system.

_____ 23. The main purpose of insulation is to support the exterior walls.

_____ 24. Outlets in homes are wired in parallel.

_____ 25. Paint is made up of pigment which gives it color and a vehicle that is the liquid portion of the paint.

NAME_____ SCHOOL_____

CLASS_____ PERIOD_____ DATE_____ SCORE_____

IDENTIFICATION: Please identify the following parts of a house:

26. _____

27. _____

28. _____

29. _____

30. _____

31. _____

32. _____

FRAMING A ROOF

33. _____

34. _____

35. _____

36. _____

37. _____

38. _____

39. _____

40. _____

41. _____

42. _____

43. _____

FRAMING AN EXTERIOR WALL

44. _____

45. _____

46. _____

47. _____

48. _____

49. _____

50. _____

PRINCIPAL TYPES OF ROOFS

BUILDING A STRUCTURE

Chapter 11, Text pages 302–341

INTERFACING WITH TECHNOLOGY

Before construction begins, a site is selected and building codes and permits are obtained. Home construction involves building the footing, foundation, floors, walls, and roof. After this, utilities and insulation are installed. When this is completed, both the inside and outside are finished. Some parts of house are built in factories. Someday you may rent or own a home. To better prepare for this, you will need to become well-informed. Answer the following questions:

SCIENCE: What do you study in science that will help you make decisions regarding the type of roof, insulation, and foundation for a house?

_____

_____

MATH: When you get a mortgage for a house, why would you need to know the effects of interest rates?

_____

_____

SOCIAL STUDIES: What has history taught us about houses and earthquakes?

_____

_____

LANGUAGE ARTS: Explain why you would need to know how to read and comprehend what you read when working with the contractor who is building or remodeling your house?

_____

_____

ART: When you are selecting a site and considering the design for the house, how do you think a background in art will help you?

_____

_____

NAME_____ SCHOOL_____

CLASS_____ PERIOD_____ DATE_____ SCORE_____

BUILDING A STRUCTURE

Chapter 11, Text pages 302–341

TEST

MULTIPLE CHOICE: Place the letter of the correct answer in the blank provided to the left of the question.

_____ 1. The preconstruction phase in building construction involves obtaining a building permit and:
   a. finishing
   b. picking a site
   c. contracting
   d. sheathing

_____ 2. Walls that are designed to support weight from above are:
   a. insulating walls
   b. load bearing walls
   c. dry walls
   d. sheathing

_____ 3. The Truss-frame system makes use of a roof truss, two wall studs, and a floor truss assembled into:
   a. one unit
   b. two units
   c. three units
   d. four units

_____ 4. Outlets in a house are wired:
   a. perpendicular
   b. in series
   c. in parallel
   d. both "B" and "C"

_____ 5. When 2 X 4 studs are used in house construction, these are spaced:
   a. 12" on center
   b. 18" on center
   c. 16" on center
   d. 26" on center

TRUE/FALSE: Write the full word TRUE or FALSE in response to the following statements.

_____ 6. The footing, foundation, floors, walls, and roof are built before utilities are installed.

_____ 7. Manufactured houses are built in a factory.

_____ 8. The purpose of insulation is to give strength to the wall and ceiling.

_____ 9. Wind effects influence the design and construction of skyscrapers.

_____ 10. Utilities include the sheathing, plumbing, and underlayment.

NAME_____ SCHOOL_____

CLASS_____ PERIOD_____ DATE_____ SCORE_____

IDENTIFICATION: Using the terms below, identify the parts for framing an exterior wall.

Stud   Bottom plate   Partition intersection
Corner post   Corner brace   Top plate
Trimmer   Header   Rough sill
Jack stud

11. _____

12. _____

13. _____

14. _____

15. _____

16. _____

17. _____

18. _____

19. _____

20. _____

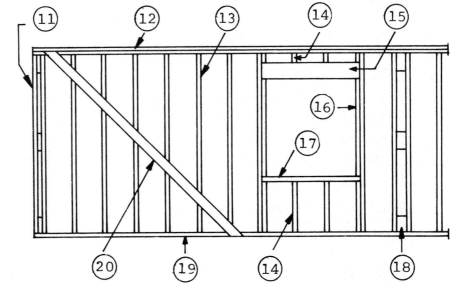

IDENTIFICATION: Below are five styles of roofing. Using the terms below, identify these roofs.

Mansard   Gambrel   Gable   Hip   Shed

21. _____

22. _____

23. _____

24. _____

25. _____

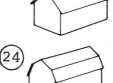

BUILDING A STRUCTURE

Chapter 11, Text pages 302–341

ANSWER SHEET

///////// KEY WORDS \\\\\\\\\

1. building permit
2. footing
3. mason
4. Mortar
5. frost line
6. header
7. sheathing
8. pitch
9. insulation
10. vapor barrier

///////// REVIEW \\\\\\\\\

FILL IN THE BLANKS:

1. a. Building the footing, foundation, floors, walls, and roof
   b. Installing utilities and insulation
   c. Finishing the structure

2. a. Picking the location   b. Obtaining building permits

3. a. Footing   b. Foundation wall

4. Commercial structures

5. National Electrical Code

MULTIPLE CHOICE:

6. B	10. C	13. C
7. A	11. C	14. B
8. D	12. A	15. D
9. A		

TRUE/FALSE:

16. True	20. False	23. False
17. False	21. False	24. True
18. False	22. True	25. True
19. True		

IDENTIFICATION:

FRAMING A ROOF

26. Ridge board
27. Collar beam
28. Rafter
29. Block
30. Top plates
31. Ceiling joist
32. End Stud

FRAMING AN EXTERIOR WALL

33. Corner post
34. Top plate
35. Stud
36. Jack stud
37. Header
38. Trimmer
39. Rough Sill
40. Partition intersection
41. Jack Stud
42. Bottom Plate
43. Corner Brace

PRINCIPAL TYPES OF ROOFS

44. Flat roof
45. Shed roof
46. Hip roof
47. Gable roof
48. Mansard roof
49. Gambrel roof
50. Deck roof

(Answers may vary and there may be additions to these suggested answers. Answers can depend upon the level of studies in the various disciplines for the group of students you are teaching.)

SCIENCE: In science, you study about weather conditions and the effects of temperatures on materials. You also study about global warming and its effects on the climate.

MATH: House mortgages can range from 9% interest to 16% interest. A variation of this can mean paying over $200.00 in interest per month for the next 30 years. This means one needs to understand the importance of only one fourth of a percent more or less interest and the increase or decrease in monthly payments.

SOCIAL STUDIES: In the past, when earthquakes shook an area, houses constructed in certain ways either withstood the tremors or fell apart. Using what we have learned in the past and what technology is available now, one can have a house built that can withstand most of the tremors of an earthquake.

LANGUAGE ARTS: If you cannot read the "fine print" on contracts, look for a person's credentials, and understand what is in the contract, you may be paying for work that will never be done. You need to be able to read and comprehend any contract before you sign it. You must also be knowledgeable enough to know which questions to ask and when to ask them.

ART: Choosing a house that fits in with the site's environment enhances the beauty of a home. Being able to visualize final landscaping takes some creativity and some basic art concepts such as shape, size, texture, and color.

//////// **TEST** \\\\\\\\

MULTIPLE CHOICE:

1. B
2. B
3. A
4. C
5. C

TRUE/FALSE:

6. True
7. True
8. False
9. True
10. False

IDENTIFICATION:

FRAMING FOR AN EXTERIOR WALL:

11. Corner post
12. Top plate
13. Stud
14. Jack stud
15. Header
16. Trimmer
17. Rough sill
18. Partition intersection
19. Bottom plate
20. Corner brace

TYPES OF ROOFS:

21. Hip roof
22. Gable roof
23. Mansard roof
24. Gambrel roof
25. Shed roof

MANAGING PRODUCTION SYSTEMS

Chapter 12, Text pages 342–370

KEY WORDS

Using the Key Words listed below, demonstrate your mastery of the Key Words by selecting the correct answer to complete each statement.

Action item	EPA	Project manager
Architect	Gantt chart	Prototype
Business plan	General contractor	Quality control
Cash flow analysis	Loss	Research and development
Certificate of occupancy	Marketing	Sales representative
Commission	OSHA	Shareholder
Contract	Overruns	Subcontractor
Coordination	Owner	Throwaway
Dealer	Permits	Union
Direct sales	PERT chart	Venture capitalist
Distributor	Productizing	Zoning
Engineer	Profit	

1. Structures are designed by _____.

2. The _____ checks to make sure the building is structurally sound.

3. The _____ _____ is a person or company who takes the overall responsibility for a construction project.

4. _____ are hired by general contractors to do specific work such as plumbing, electricity, carpentry, or masonry.

5. Additional costs created by items being left out of bids or poor management are called cost _____.

6. On very large projects, _____ _____ oversee contracts, scheduling, material deliveries, and overall progress of any job.

7. _____ are labor organizations that work with companies to set pay and working conditions for their members.

8. When issues cannot be resolved in project meetings, these _____ _____ are recorded and brought up at the next meeting along with a report of the work done on each action item.

9. The _____ _____ is used to help managers schedule and monitor a project.

10. A/An _____ _____ _____ is issued after all systems are checked and approved and the building is finished.

11. The _____ department decides who the customers are, what products to make, and how much the project will cost.

12. _____ and _____ is the search for new materials, processes, and methods and the use of these discoveries.

13. A/An _____ is a test version of the product.

14. A/An _____ is a product that cannot be fixed and must be replaced.

15. In _____ _____ the company salespeople sell directly to customers.

16. A/An _____ _____ is a person who sells products for a company when it does not have enough sales to open an office.

17. The percentage of the sales price that the manufacturer pays the sales representative is called a/an _____.

18. The _____ sells products to people by taking orders for them and placing them with the manufacturer.

19. The _____ _____ _____ predicts how much money will have to be spent each week or each month and how much income will come in during that time period.

20. A/An _____ _____ is a person who supplies money to a new or growing company and in return receives part ownership in the company.

NAME_____ SCHOOL_____

CLASS_____ PERIOD_____ DATE_____ SCORE_____

MANAGING PRODUCTION SYSTEMS

Chapter 12, Text pages 342–370

REVIEW

MATCHING: Below are various terms used in manufacturing. Place the letter of the correct response in the blank to the left of term it matches.

_____ 1. Contract	A.	scheduling technique developed by Navy
_____ 2. Market	B.	product that cannot be repaired
_____ 3. Research	C.	a percentage of the sale
_____ 4. Commission	D.	sets pay and working conditions
_____ 5. Union	E.	search for new materials and products
_____ 6. Life safety	F.	agreement between owner and builder
_____ 7. Prototype	G.	potential customers
_____ 8. PERT	H.	plan that tells how stairs and exits will protect people in case of fire
_____ 9. Gantt	I.	test version of a product
_____10. Throwaway	J.	bar graph used to help managers schedule and monitor a project

TRUE/FALSE: Write the full word TRUE or FALSE in response to the following statements:

_____ 11. The coordination and proper use of the seven resources of technology used in the manufacturing or construction system are part of the job of a company's management.

_____ 12. An architect is the person who accepts complete responsibility for a construction project.

_____ 13. A contract is the agreement between the owner and the general contractor.

_____ 14. Subcontracting is when the general contractor has his own employees to do the entire job.

_____ 15. Unions are labor organizations that bargain with employers to set wages and work practices for their members.

_____ 16. Permits and approvals are designed to protect the safety of workers and the public near projects.

_____ 17. A life safety plan in a construction project is the insurance policy the union has for members.

_____ 18. The Certificate of Occupancy or CO is issued when all systems have been tested and the building is complete.

_____ 19. A jurisdiction is the carefully-described portion of work claimed by the union.

_____ 20. Zoning exists when industry is allowed to move into residential areas.

MULTIPLE CHOICE: Place the letter of the correct answer in the blank provided to the left of each question.

_____21. The marketing department identifies potential customers who are called the:
    a.   market gap analysis   c.   market
    b.   market share   d.   total market

_____22. The amount of a product bought in one year is the:
    a.   Market gap analysis   c.   market
    b.   Market share   d.   total market

_____23. The use of new discoveries and previously known methods and techniques to solve a specific problem is:
    a.   Development   c.   market gap analysis
    b.   Research   d.   production

_____24. The repair and maintenance of a product once it is marketed is known as:
    a.   Development   c.   servicing
    b.   Productizing   d.   both "b" and "c"

_____25. When a product is out of warranty and is no longer repairable, it is known as a:
    a.   Repairable product   c.   throwaway
    b.   Cash flow   d.   warranty

FILL IN THE BLANKS: In the blanks provided, write the correct answer to each question.

26. _____ _____ refers to having salespeople who are employees of the company selling products directly to customers.

27. The_____ states the goals of the business, the strategy and methods that will be used to reach the goals, and amount of money needed to start the business and keep it going.

28. The_____ predicts how much money will have to be spent each week or month of operation and how much income is expected in the same periods.

29. A _____ _____ supplies money to a new or growing company in exchange for part ownership in the company.

30. _____ includes planning, organizing, leading, and controlling.

MANAGING PRODUCTION SYSTEMS

Chapter 12, Text pages 342–370

INTERFACING WITH TECHNOLOGY

The success of a production system depends on its management. The seven resources must be carefully coordinated and used. Management must coordinate the work of all the departments. Marketing decides what market, the features, and the way to sell products. We as consumers and later as a part of the production system must have a general background to help us make wise choices. Let us determine how each of the subjects we study affects us.

SCIENCE: What do you study in science that will help you make decisions regarding the purchase of products?

_____

_____

MATH: What is used in managing production systems that relates to math?

_____

_____

SOCIAL STUDIES: What do you think is happening now that will become an important piece of information that will be recorded in history?

_____

_____

LANGUAGE ARTS: Where in the production system does reading become very useful?

_____

_____

ART: If you are in research and development, what would you research about a product that relates to art?

_____

_____

MANAGING PRODUCTION SYSTEMS

Chapter 12, Text pages 342-370

TEST

MULTIPLE CHOICE: Place the letter of the correct answer in the blank provided to the left of each question.

_____ 1. The group that researches and develops products is known as:
    a. R & D   c. QC
    b. OSHA   d. HVAC

_____ 2. The agreement between the general contractor and the owner is called a:
    a. cash flow analysis   c. contract
    b. commission   d. jurisdiction

_____ 3. The search for new materials, processes, and techniques that may be useful in a company's business is called:
    a. development   c. research
    b. productizing   d. prototyping

_____ 4. A kind of bar graph that helps managers schedule and monitor a project showing the beginning and end is called:
    a. PERT chart   c. life safety plan
    b. Gantt chart   d. business plan

_____ 5. That which predicts how much money will have to be spent each week or month of operation and how much income can be expected is called the:
    a. market   c. cash flow analysis
    b. direct sales   d. venture capitalist

TRUE/FALSE: Write the full word TRUE or FALSE in response to the following statements:

_____ 6. Subcontracting is when the general contractor hires his own employees to do the entire job.

_____ 7. When a product is out of warranty it may be regarded as a throwaway.

_____ 8. The Certificate of Occupancy is issued when all systems have been tested and the building is complete.

_____ 9. Architects design a building's structure and its major systems.

_____ 10. The coordination and proper use of the seven resources of technology used in the manufacturing or construction system are part of the job of a company's management.

MANAGING PRODUCTION SYSTEMS

Chapter 12, Text pages 342–370

ANSWER SHEET

////////// KEY WORDS \\\\\\\\\\

1. architects
2. engineer
3. general contractor
4. Subcontractors
5. overruns
6. project managers
7. Unions
8. action items
9. Gantt Chart
10. Certificate of Occupancy
11. marketing
12. Research, development
13. prototype
14. throwaway
15. direct sales
16. sales representative
17. commission
18. dealer
19. cash flow analysis
20. venture capitalist

//////// REVIEW \\\\\\\\\

MATCHING:

1. F   6. H
2. G   7. I
3. E   8. A
4. C   9. J
5. D   10. B

TRUE/FALSE:

11. True   16. False
12. False   17. False
13. True   18. True
14. False   19. True
15. True   20. False

MULTIPLE CHOICE:

21. C
22. D
23. A
24. C
25. C

FILL IN THE BLANKS:

26. Direct sales
27. business plan
28. cash flow analysis
29. venture capitalist
30. Managing

//////// INTERFACING WITH TECHNOLOGY \\\\\\\\\

(Answers may vary and there may be additions to these suggested answers. Answers can depend upon the level of studies in the various disciplines for the group of students you are teaching.)

SCIENCE: In science we study the effects of using our resources on the environment. Knowledge of the impact of overusing a resource could cause us to not purchase certain products or participate in a recycling effort.

MATH: When managing production systems, the use of Gantt and PERT charts help schedule and monitor. In math we learn how to make, read, and understand charts.

SOCIAL STUDIES: History will show the great impact computers had on managing production systems – the changes made in output of kinds and amounts of computer-generated information in all phases of managing a production system.

LANGUAGE ARTS: Reading is very useful when it comes to reading all the contracts and specifications.

ART: In research and development, you would do studies relating to designs preferred by the consumer. In development, you would work with the designs and develop a design that is compatible to the usefulness and materials used in construction of the product.

///////// *TEST* \\\\\\\\\\

MULTIPLE CHOICE:

1. A
2. C
3. C
4. B
5. C

TRUE/FALSE:

6. False
7. True
8. True
9. False
10. True

ENERGY

Chapter 13, Text pages 372–399

KEY WORDS

Using the Key Words listed below, demonstrate your mastery of the Key Words by selecting the correct answer to complete each statement.

Active solar	Fusion	Passive solar
Biomass	Gasohol	Potential energy
Conservation of energy	Gasification	Quad
Energy	Hydroelectricity	Radiation
Fermentation	Kinetic energy	Solar cell
Fission	Nuclear fuel	Work
Fossil fuel	Parabolic reflector	

1. _____ is done when a force pushes or pulls on an object, causing the object to move.

2. _____ is the ability to do work.

3. The energy that is stored in an object because of its position, shape, or some other feature is called _____ _____.

4. The energy an object has because of its motion is called _____ _____.

5. Coal, oil, and gas are known as _____ _____.

6. The rotting of biomass produces methane gas in a process called_____.

7. A _____ is one quadrillion British Thermal Units.

8. In nuclear _____, a large atom such as uranium-235 is hit with tiny particles called neutrons. The split results in energy given off in the form of heat and light.

9. Energy produced by falling water turning turbines is called_____.

10. A biomass process called fermentation produces _____, which is a mixture of gasoline and alcohol.

ENERGY

Chapter 13, Text pages 372–399

REVIEW

TERMINOLOGY: Please define or explain the following terms:.

1. Work _____.

2. Kinetic energy is _____
    _____.

3. Potential energy is _____
    _____.

4. The principle of conservation of energy states _____
    _____
    _____.

5. A quad is _____.

6. A British Thermal Unit (BTU) is _____
    _____.

FILL IN THE BLANKS WITH THE INFORMATION THAT IS MISSING.

TYPES OF ENERGY SOURCES

LIMITED	UNLIMITED	RENEWABLE
7. _____	11. _____	17. _____
8. _____	12. _____	18. _____
9. _____	13. _____	19. _____
10. _____	14. _____	20. _____
	15. _____	21. _____
	16. _____	

MULTIPLE CHOICE: Place the letter of the correct answer in the blank provided to the left of each question.

_____22. As energy sources, oil, natural gas, and coal are all called:
    a.   fossil fuels   c.   unlimited sources
    b.   limited sources   d.   both "a" and "b"

_____23. The energy source we depend on the most is:
    a.   nuclear fuel   c.   oil
    b.   coal   d.   gas

_____24. The instrument that measures vibrations in the earth created by setting off a small explosive just beneath the surface of the earth is a:
    a.   magnetometer   c.   gravimeter
    b.   seismograph   d.   geiger counter

_____25. Fuel obtained by bombarding Uranium-235 with neutrons is called:
    a.   fission   c.   fusion
    b.   photovoltaic cells   d.   gasification

TRUE/FALSE: Write the full word TRUE or FALSE in response to the following statements.

_____ 26. An example of chemical energy would be the burning of fossil fuels.

_____ 27. Examples of unlimited sources of energy include the sun, wind, and tides.

_____ 28. Both gas and coal can be pumped through pipe lines.

_____ 29. The nuclear wastes from nuclear fission plants give off high-energy particles called radiation that takes up to a hundred years before it is no longer radioactive.

_____ 30. The sun, moon, and other planets' gravity cause wind energy.

_____ 31. Biomass is a renewable source of energy created by the processes of direct combustion, gasification, or fermentation.

_____ 32. Geothermal energy is created by heat from the center of the earth.

_____ 33. The earliest sources of energy were human and animal muscle power.

ENERGY

Chapter 13, Text pages 372–399

INTERFACING WITH TECHNOLOGY

Energy can be kinetic or potential. It is also classified as limited, unlimited, and renewable. Over half the energy we use comes from oil. The rest comes from coal, gas, wind, the sun, nuclear, and geothermal energy. Most of our present sources of energy are being used up. How can a knowledge in the different subject areas prepare us as consumers?

SCIENCE: What do you study in science about nuclear energy?

_____

_____

MATH: If we received a bill each month for our use of electricity, what can we tell from reading the numbers on the bill?

_____

_____

SOCIAL STUDIES: Why is it that just 100 years ago we used so little energy in comparison to the present?

_____

_____

LANGUAGE ARTS: Being able to read and comprehend what is written in newspapers and magazines helps us. What is one item about sources of energy that makes big news?

_____

_____

ART: If you worked for a utility company such as the gas or electric company, how might a degree in art help you get a job there?

_____

_____

NAME_____ SCHOOL_____

CLASS_____ PERIOD_____ DATE_____ SCORE_____

ENERGY

Chapter 13, Text pages 372–399

TEST

TRUE/FALSE: Write the full word TRUE or FALSE in response to the following statements.

_____ 1. Limited sources of fuels such as coal, oil, and gas are known as fossil fuels.

_____ 2. Nuclear energy provides an unlimited and safe form of energy for the future.

_____ 3. Unlimited sources of energy such as wind and tidal energy come from the sun.

_____ 4. Examples of renewable energy sources include biomass, wood, and human and animal muscle power.

_____ 5. In the search for oil and gas, magnetometers, seismographs, and photovoltaic cells are used.

MULTIPLE CHOICE: Place the letter of the correct answer in the blank provided to the left of each question.

_____ 6. Radioactivity from nuclear wastes can last up to:
    a. thousands of years   c. hundreds of years
    b. 100 years   d. one million years

_____ 7. Energy in motion is known as:
    a. potential   c. kinetic
    b. photocells   d. neutrons

_____ 8. The energy we depend on the most is:
    a. coal   c. nuclear
    b. geothermal   d. oil

_____ 9. Unlimited sources of energy include wind and:
    a. solar   c. fusion
    b. geothermal   d. "a," "b," & "c"

_____10. Limited sources of energy include coal, oil, and:
    a. tidal   c. biomass gasification
    b. uranium   d. fusion

ENERGY

ANSWER SHEET

///////// KEY WORDS \\\\\\\\\\

1. Energy
2. Work
3. potential energy
4. kinetic energy
5. fossil fuels

6. gasification
7. quad
8. fission
9. hydroelectricity
10. gasohol

///////// REVIEW \\\\\\\\\\

TERMINOLOGY:

1. . . .is equal to the distance an object moves, multiplied by the force in the direction of the motion.

2. . . .energy in motion

3. . . .energy that is stored in an object due to its position, shape, or other feature.

4. . . .that energy cannot be created or destroyed, but only changed from one form to another.

5. . . .one quadrillion BTUs, which is equal to the energy given off by burning 20 million gallons of gasoline every day for a year.

6. . . .is the amount of energy needed to raise the temperature of one pound of water by one degree Fahrenheit.

FILL IN THE BLANKS:

7. Coal
8. Oil
9. Natural gas
10. Uranium

11. Solar
12. Wind
13. Gravitational
14. Tidal
15. Geothermal
16. Fusion

17. Wood
18. Biomass gasification
19. Biomass Fermentation
20. Animal Power
21. Human Muscle Power

MULTIPLE CHOICE:

22. D
23. C
24. B
25. A

TRUE/FALSE:

26. True
27. True
28. True
29. False
30. False
31. True
32. True
33. True

///////// INTERFACING WITH TECHNOLOGY \\\\\\\\\

(Answers may vary and there may be additions to these suggested answers. Answers can depend upon the level of studies in the various disciplines for the group of students you are teaching.)

SCIENCE: In technology, we learn how uranium-235 is used to create nuclear energy. In science, we study about uranium as an element. We also learn about radiation and the waste material of the process of fission called radioactive waste. In technology, people find ways to bury the radioactive waste.

MATH: When we get our electrical bill, we can find how many kilowatt hours we have used. We can also determine the cost per hour of electricity and the total cost.

SOCIAL STUDIES: There were not as many people 100 years ago. Also, many of today's technologies and our economy allows us to use many devices and conveniences that require much more electricity to run.

LANGUAGE ARTS: One item about energy that makes big headlines is the depletion of our energy sources. Another item that makes worldwide news is when a nuclear power plant accident occurs. If we can read and understand what we read, we will be more likely to make wise use of our energy sources. We will also take responsibility to become an informed voter when faced with issues relating to energy.

ART: Utility companies produce brochures relating to their product. A degree in art will help you find a position such as layout artist.

///////// T E S T \\\\\\\\\

TRUE/FALSE:
1. True
2. False
3. True
4. True
5. False

MULTIPLE CHOICE:
6. A
7. C
8. D
9. D
10. B

POWER

Chapter 14, Text pages 400–423

KEY WORDS

Using the Key Words listed below, demonstrate your mastery of the Key Words by selecting the correct answer to complete each statement.

Absolute zero	Horsepower	Power system
Alternating current	Hydraulic	Pressure
Battery	Idler wheel	Reaction engine
Diesel engine	Internal combustion engine	Rocket engine
Electric motor	Jet engine	Rotor
Engine	Load	Stator
External combustion	Momentum	Superconductor
engine	Nuclear reactor	Transformer
Four-stroke cycle	Pneumatic	Transmission
Fractional horsepower	Poles	Two-stroke cycle
Generator	Power	Watt

1. The amount of work done during a given period of time is called _____.

2. The measurement for power is the _____.

3. A/An _____ is a device that transfers force from one place to another or changes its direction.

4. In an _____ _____ _____, fuel is burned in one chamber, heating liquid or gas in another chamber causing expansion and an increase in pressure. This applies a force to a piston or turbine which then moves.

5. The _____ _____ _____ is used in airplanes, cars, trucks, ships, and railroad locomotives.

6. The _____ _____ and the _____ _____ both operate on Newton's third law of motion: for every action, there is an equal and opposite reaction.

7. A/An _____ is a device that is connected to a turbine, which changes rotary motion into electrical energy.

8. In a generator, _____, which has large permanent magnets attached to it, spins within a non-moving housing called a _____.

9. Motors inside toys, appliances, and small machines which have less than one horsepower are called _____ _____.

10. _____ is the force on a liquid or object divided by the area over which it is applied.

11. _____ _____ is produced by generator plants.

12. _____ are used to change the high voltage to lower voltage near the place where the electricity is used.

13. A _____ is a material that has no electrical resistance.

14. The temperature at which molecules stop moving (about 459 degrees Fahrenheit) is called _____ _____.

15. A battery can be recharged by a generator called a/an _____.

16. A/An _____ is used to store and change chemical energy into electrical energy.

17. In a _____ _____, fission occurs which produces heat that boils water, making steam that turns a turbine.

18. _____ occurs when a flywheel begins moving and keeps on turning.

19. A/An_____ is a machine that uses energy to create mechanical force and motion.

20. A/An _____ is a unit of measurement in the metric system which measures power.

POWER

Chapter 14, Text pages 400–423

REVIEW

FILL IN THE BLANKS: In the blanks provided, write the correct answer to each question.

1. Power is_____

   _____.

2. An engine is_____

   _____.

3. A transmission is_____

   _____.

4. Newton's third law of motion states_____

   _____.

5. Five types of transmissions are:

   a. _____   c. _____   e. _____

   b. _____   d. _____

MATCHING: Match the following words with the correct definition. Put the letter next to the correct statement in the blank.

_____ 6. Reaction engine
_____ 7. Generator
_____ 8. Nuclear reactor
_____ 9. Pressure
_____10. Horsepower
_____11. Jet engine
_____12. Electric motor
_____13. Rocket engine
_____14. Transformer

a. machine that converts rotary motion into electricity
b. it is equal to 550 foot-pounds per second
c. makes energy by splitting large atomic nuclei
d. force on a liquid or object divided by the area over which it's applied
e. air is pushed into a combustion chamber by a compressor in it
f. used to convert high voltage current to a lower voltage
g. carries its own supply of oxygen
h. converts electrical energy into rotary motion
i. examples are the jet and rocket engines

MULTIPLE CHOICE: Place the letter of the correct answer in the blank provided for each question.

_____15. In this engine, fuel is burned in one chamber, heating liquid or a gas in another chamber:

   a. internal combustion   c. external combustion
   b. reaction   d. jet

____16. In this engine, fuel is made to explode in a chamber, pushing a piston which is connected to a crankshaft. This is connected to a crankshaft and flywheel and causes rotary motion:
   a. internal combustion   c. external combustion
   b. reaction   d. rocket

____17. In this engine, fuel is mixed with air, creating a rapidly expanding hot exhaust gas:
   a. internal combustion   c. external combustion
   b. reaction   d. machine

____18. Simple parts such as gears, cams, levers, pulleys, and linkages are used in what type of transmission system?
   a. electrical   c. mechanical
   b. pneumatic   d. magnetic

____19. Water or other liquids are used in what type of transmission system?
   a. electrical   c. hydraulic
   b. pneumatic   d. magnetic

____20. Electrical current is used in what transmission system?
   a. mechanical   c. hydraulic
   b. pneumatic   d. electrical

TRUE/FALSE: Write the full word TRUE or FALSE in response to the statements.

_____ 21. Fractional horsepower motors deliver less than one horsepower.

_____ 22. Alternating current is produced by generators.

_____ 23. A superconductor is the type of wire widely used to carry electricity to residential areas.

_____ 24. Absolute zero is the temperature at which all molecular motion stops.

_____ 25. Batteries change chemical energy into electrical energy.

_____ 26. Pressure is the amount of work done in a given period of time.

POWER

Chapter 14, Text pages 400-423

INTERFACING WITH TECHNOLOGY

Power systems have two parts: an engine and a transmission. An engine is a machine that uses energy to create mechanical force and motion. The transmission is a device that transfers force from one place to another. Engines can be internal or external combustion engines, reaction engines, electric motors, and nuclear reactors. Since we make use of power systems every day, we need to learn more about them. How does each of these subjects help you learn more about power systems?

SCIENCE: What do you study in science about power systems?

_____

_____

MATH: How will math help us if we are using or purchasing an engine or a device that has an engine?

_____

_____

SOCIAL STUDIES: What power system was used during the Industrial Revolution and how was it used?

_____

_____

LANGUAGE ARTS: Being able to read and comprehend what is written is important. Why is it so important when you are using or working with engines?

_____

_____

ART: How would a nuclear power company use a person who has a commercial art degree?

_____

_____

POWER

Chapter 14, Text pages 400–423

TEST

TRUE/FALSE: Write the full word TRUE or FALSE in response to the statements.

_____ 1. In an external combustion engine, fuel is burned in one chamber, heating liquid or a gas in another.

_____ 2. Gears, cams, levers, and pulleys are used in mechanical transmissions.

_____ 3. Batteries convert electrical energy into chemical energy.

_____ 4. A lawn mower engine is an example of a reaction engine.

_____ 5. Pneumatic transmissions systems use air pressure.

_____ 6. Newton's third law of motion states that for every action, there is an equal and opposite reaction.

_____ 7. Fractional horsepower motors deliver more than one horsepower.

_____ 8. A transformer is a device that pushes air into a combustion chamber to create a reaction.

FILL IN THE BLANKS: In the blanks provided, write the correct answer to each question.

9. An _____ motor converts electrical energy into rotary motion.

10. _____ is the amount of work done during a given time.

11. A _____ is a device that transfers force from one place to another or changes direction.

12. _____ engines are a form of reaction engine that carries its own supply of oxygen.

13. _____ are types of wire that carry electrical current with no resistance.

14. _____ _____ is the temperature at which all molecular motion stops.

15. One _____ is equal to 550 foot-pounds per second.

POWER

Chapter 14, Text pages 400–423

ANSWER SHEET
///////// KEY WORDS \\\\\\\\\

1. power
2. horsepower
3. transmission
4. external combustion engine
5. internal combustion engine
6. jet engine, rocket engine
7. generator
8. rotor, stator
9. fractional horsepower
10. Pressure

11. Alternating current
12. Transformers
13. superconductor
14. absolute zero
15. alternator
16. battery
17. nuclear reactor
18. Momentum
19. engine
20. watt

///////// REVIEW \\\\\\\\\

FILL IN THE BLANKS:

1. ...the amount of work done during a given period of time.

2. ...a machine that uses energy to create mechanical force and motion.

3. ...a device that transfers force from one place to another or changes its direction.

4. ...that for every action, there is an equal and opposite reaction.

5. a. Mechanical
   b. Hydraulic (fluid)
   c. Pneumatic (air pressure)
   d. Electrical
   e. Magnetic

MATCHING:	MULTIPLE CHOICE:	TRUE/FALSE:
6. I	15. c	21. True
7. A	16. a	22. True
8. C	17. b	23. False
9. D	18. c	24. True
10. B	19. c	25. True
11. E	20. d	26. False
12. H		
13. G		
14. F		

///////// INTERFACING WITH TECHNOLOGY \\\\\\\\\

(Answers may vary and there may be additions to these suggested answers. Answers can depend upon the level of studies in the various disciplines for the group of students you are teaching.)

SCIENCE: In science we study about watts, horsepower, pressure, and power. In technology we use the devices about which we study in science. We also study about the conversion of a

347

liquid to a gas which forms pressure inside a vessel. In technology we make use of the conversion of liquid to a gas for the purpose of creating energy.

MATH: It is good to understand numbers and the scientific concepts behind the rating of engines and motors. We need to know how much energy we need and for how long.

SOCIAL STUDIES: The steam engine was used during the Industrial Revolution to run the machines in the factories.

LANGUAGE ARTS: Any time that we try to learn to repair or maintain an engine, we need to be able to read and understand repair and/or instruction manuals.

ART: Nuclear power plants are highly criticized for their production of very dangerous radioactive wastes. To inform and persuade the public to accept their existence, large quantities of public relations materials are produced and distributed. Commercial artists design these pieces of literature.

////////// **TEST** \\\\\\\\\\

TRUE/FALSE:

1. True
2. True
3. False
4. False
5. True
6. True
7. False
8. False

FILL IN THE BLANKS:

9. electric
10. Power
11. transmission
12. Rocket
13. Superconductors
14. Absolute zero
15. horsepower

NAME_____ SCHOOL_____

CLASS_____ PERIOD_____ DATE_____ SCORE_____

TRANSPORTATION

Chapter 15, Text pages 424–455

KEY WORDS

Using the Key Words listed below, demonstrate your mastery of the Key Words by selecting the correct answer to complete each statement.

Buoyancy	Engine	Lift	Thrust
Commute	Intermodal	Piggyback	Transmission
Container	Internal	Pipeline	Turbo-prop
ship	combustion	Propeller	Vehicle
Diesel	engine	Rocket	Weight
Drag	Jet	Steam engine	

1. A/An _____ transportation system is made up of several kinds of transportation moving goods over long distances.

2. The gears or belts that connect the motor to the wheels are called the _____ or drive.

3. The _____ _____ _____ replaced the steam engine with its smaller size and greater fuel economy.

4. A_____ engine does not use spark plugs, but relies on compression of the fuel in which the fuel becomes very hot and explodes with a spark.

5. An object placed into the water is pushed upward by a force equal to the weight of the object, which in turn creates _____.

6. Transportation vehicles that carry over the water large boxes that can be attached to trucks or placed on railroad cars are called _____ _____.

7. The carrying of tractor trailers by trains is called _____.

8. The four forces acting on an airplane are _____, _____, _____, and _____.

9. The _____ of an airplane move the air from front to back, pushing the airplane through the air.

10. Jet engines use oxygen in the air to burn fuel, whereas a _____ must carry its own oxygen. This allows it to travel in high altitudes.

TRANSPORTATION

Chapter 15, Text pages 424–455

REVIEW

MATCHING: Below are people who are well known for their contributions to transportation technology. Match the contribution to the person and place the letter of the response in the blank provided.

_____ 1. Westinghouse
_____ 2. Nicholas Otto
_____ 3. Robert Fulton
_____ 4. Chuck Yeager
_____ 5. Yuri Gagarin
_____ 6. Neil Armstrong
        & Edwin Aldrin
_____ 7. Wright brothers
_____ 8. Jacques Cousteau
_____ 9. Nicolas-Joseph Cugnot

A. First exploration of the moon
B. First manned flight in space
C. First sustained powered flight
D. Invented the air brake
E. Built a modern ship with wind power
F. Invented the internal combustion engine
G. Built the first vehicle to be moved by mechanical power (1769)
H. Built the first steam-powered ship to be used on a regular basis
I. First person to fly at more than the speed of sound

MULTIPLE CHOICE: The following questions relate to resources for transportation systems. Place the letter of the correct answer in the blank provided to the left of each question.

_____10. Aluminum, composites, and titanium are which resource for transportation systems?
    a. Information   c. Materials
    b. Tools and machines   d. Capital

_____11. Conveyor belts, pipelines, and vehicles are which resource for transportation systems?
    a. Information   c. Materials
    b. Tools and machines   d. Capital

_____12. Routes, speed and position, and operation of vehicles are which resource for transportation systems?
    a. Information   c. Time
    b. Capital   d. People

_____13. Money collected through tolls or taxes to build and maintain ports or roadways is which resource for transportation systems?
    a. Information   c. Time
    b. Capital   d. People

_____14. Motormen, pilots, astronauts, and drivers refer to which resource for transportation systems?
    a. Information   c. Time
    b. Capital   d. People

_____15. Human power, wind, electricity, and gasoline refer to which resource for transportation systems?
   a. People   c. Energy
   b. Tools and machines   d. Information

TRUE/FALSE: Write the full word TRUE or FALSE in response to the following statements.

_____ 16. Modern transportation and communication systems have helped to make countries interdependent.

_____ 17. Objects float if their weight is more than their buoyancy.

_____ 18. Ships that operate below the surface of water are called submersibles or submarines.

_____ 19. Intermodal transportation systems make the best use of each type of transportation system, but they involve excessive handling and expenses.

_____ 20. Lighter-than-air (LTA) vehicles use passive lift, which means these float in air.

_____ 21. To escape the Earth's gravity, a speed of close to Mach 1 must be reached.

_____ 22. Examples of nonvehicle forms of transportation include pipelines, conveyor belts, and trucks.

_____ 23. One form of a people mover is the PRT, or personal rapid transit system.

_____ 24. While elevators move people vertically, PRTs move people horizontally.

_____ 25. Transportation systems convert motion into energy.

TRANSPORTATION
Chapter 15, Text pages 424–455

INTERFACING WITH TECHNOLOGY

Transportation has made it possible for us to exchange goods and ideas and travel more often and to further distances. We can travel by land, sea, air, space, and nonvehicle systems. How will a good background in all the subjects we study help us in using modern transportation systems?

SCIENCE: What is learned in science about traveling in the air and space?

_____

_____

MATH: If we buy a transportation vehicle or pay to travel on one, how is math used?

_____

_____

SOCIAL STUDIES: What type transportation systems were used before the wheel was invented?

_____

_____

LANGUAGE ARTS: Being able to read and comprehend what is written is important. Why is it so important when you are planning a trip?

_____

_____

ART: How will what you learn in art influence you in the purchase of a vehicle?

_____

_____

TRANSPORTATION

Chapter 15, Text pages 424–455

TEST

MULTIPLE CHOICE: Place the letter of the correct answer in the blank provided.

_____ 1. He built the first vehicle to be moved by mechanical power:
   a. Westinghouse   c. Yuri Gagarin
   b. Robert Fulton   d. Nicolas-Joseph Cugnot

_____ 2. He was the first person to fly at more than the speed of sound:
   a. Yuri Gagarin   c. Chuck Yeager
   b. Nicholas Otto   d. George Westinghouse

_____ 3. Human power, wind, electricity, and gasoline refer to which resource for transportation systems?
   a. People   c. Materials
   b. Energy   d. Capital

_____ 4. Objects float if their buoyancy is more than their:
   a. Size   c. Weight
   b. Length and width   d. Both "a" and "b"

_____ 5. Lighter-than-air (LTA) vehicles use what type of lift?
   a. Passive   c. Active
   b. Hydraulic   d. Both "c" and "b"

TRUE/FALSE: Write the full word TRUE or FALSE in response to the following statements.

_____ 6. Intermodal transportation systems make optimum use of each type of transportation used in the system.

_____ 7. Transportation systems convert energy into motion.

_____ 8. To escape Earth's gravity, a speed of more than 17,000 miles per hour must be reached.

_____ 9. Examples of nonvehicle forms of transportation include pipelines and conveyor belts.

_____ 10. Chuck Yeager made the first manned flight into space.

TRANSPORTATION

Chapter 15, Text pages 424–455

ANSWER SHEET

/////////// KEY WORDS \\\\\\\\\\\

1. intermodal
2. transmission
3. internal combustion engine
4. diesel
5. buoyancy
6. container ships
7. piggyback
8. drag, lift, thrust, weight
9. propellers
10. rocket

/////////// REVIEW \\\\\\\\\\\

MATCHING:				MULTIPLE CHOICE:				TRUE/FALSE:			
1. D		6. A		10. C		13. B		16. True		21. False	
2. F		7. C		11. B		14. D		17. False		22. False	
3. H		8. E		12. A		15. C		18. True		23. True	
4. I		9. G						19. False		24. True	
5. B								20. True		25. False	

/////////// INTERFACING WITH TECHNOLOGY \\\\\\\\\\\

(Answers may vary and there may be additions to these suggested answers. Answers can depend upon the level of studies in the various disciplines for the group of students you are teaching.)

SCIENCE: In science we learn about atmospheric conditions and gravity. In technology we develop ways to detect and overcome atmospheric conditions and gravity.

MATH: When you are buying a vehicle, you will need to understand the effect of interest rates on the overall costs. You will also need to be able to mentally calculate what you can afford and what you feel is a fair price for a vehicle. This means comparing costs in advertisements and looking in magazines and books for costs. When traveling, you will need to be able to work with traveler's checks and credit cards. Being able to keep accurate track of expenditures will keep you from overusing your finances.

SOCIAL STUDIES: Before the wheel was invented, animals were used to carry items on sleds or on their backs.

LANGUAGE ARTS: When you are planning a trip, the ability to read and comprehend enables you to find as much information as possible and select the places you wish to go without making time-consuming mistakes. Reading about places enables you to choose the type of transportation you will use for business or vacation travel.

ART: In art you will learn about different colors and textures. You will develop a skill in recognizing good design and color combinations. The artistic tastes you develop then will help you choose the style and color of car.

/////////// TEST \\\\\\\\\\\

MULTIPLE CHOICE:

1. d
2. c
3. b
4. c
5. a

TRUE/FALSE:

6. True
7. True
8. True
9. True
10. False

IMPACTS FOR TODAY AND TOMORROW

Chapter 16, Text pages 464–493

KEY WORDS

Using the Key Words listed below, demonstrate your mastery of the Key Words by selecting the correct answer to complete each statement.

Acid rain	Growth hormone	Personal rapid transit (PRT)
Confluence	Halophyte	Pollution
Delphi survey	Impact	Smart houses
Environment	Intelligent	Speech synthesis
Ergonomics	buildings	Sulfur dioxide
Fusion	Magnetic levitation	Trend analysis
Futures wheel	(Maglev)	Videoconference
Futuring	Manufacturing in space	Voice recognition

1. Pollution in the air falls back to the earth in the form of _____ _____.

2. The technology that makes possible computers speaking like humans is called _____ _____.

3. A computer responding to the sound of the human voice is called _____ _____.

4. A/An_____ is a way of meeting different people in different places by television screen.

5. The _____ _____ is a way of forecasting the future using an idea in the center of a large circle. These central ideas lead to other ideas which in turn can lead to other ideas.

6. A/An _____ _____ is a way of forecasting the future experts use by listing ten future possibilities. Lists of their ideas are arranged and combined and ranked.

7. A/An _____ _____ is a way of forecasting the future by looking at the past.

8. _____ _____ _____ are automated vehicles that travel on tracks or guideways either above or below the ground.

9. _____ _____ will have voice-activated television and radio where people can live and work at home.

10. _____ _____ will have a central computer tying all offices together. The security and energy systems will be computerized.

IMPACTS FOR TODAY AND TOMORROW

Chapter 16, Text pages 464–493

REVIEW

FILL IN THE BLANKS: In the blanks provided, write the correct answer to each question.

1. Four possible combinations of outputs from technological systems are:

    a. _____   c. _____

    b. _____   d. _____

2. People can influence the development of technology by:

    a. _____

    b. _____

3. Four techniques of forecasting the future are:

    a. _____

    b. _____

    c. _____

    d. _____

4. People determine whether technology is a good or bad by the

    _____.

IDENTIFICATION: The following are abbreviations for technological systems. Write the meaning of each abbreviation.

5. NRC_____

6. HDTV_____

7. ISF_____

8. POTS_____

9. LTA_____

10. PRT_____

NAME_____ SCHOOL_____

CLASS_____ PERIOD_____ DATE_____ SCORE_____

MATCHING: Place the letter of the correct response in the blank to the left of the term it matches.

_____11. Pollution
_____12. Ergonomics
_____13. Sulfur dioxide
_____14. Magnus LTA
_____15. Maglev train
_____16. Halophytes
_____17. Fusion
_____18. Futurist

A. Spherical airship capable of carrying over 50 tons.
B. Vehicle equipped with strong magnets that floats above its track.
C. Predict what technology will be like in the future.
D. Applies technology to the physical needs of human beings.
E. Carbon monoxide we breathe and wastes dumped into bodies of water.
F. Technology that joins atoms together, creating a new form of matter.
G. Plants that can grow in salt water and are edible.
H. Pollutant of a coal-fired plant.

TRUE/FALSE: Write the full word TRUE or FALSE in response to the following statements.

_____ 19. Technology assessments studies determine which companies are allowed to pollute.

_____ 20. Speech synthesis is the computer technology which enables the computer to produce human voice sounds.

_____ 21. Videoconferencing is the technology allowing people to hold face-to-face meetings by satellite with people in other parts of the world.

_____ 22. In space, molecules of materials tend to hold together more than on earth.

_____ 23. The flowing together of technology systems is called confluence.

_____ 24. A Delphi survey is a technological tool used for satellite population studies.

_____ 25. Using futuring techniques, people can anticipate the consequences of a new technology.

IMPACTS FOR TODAY AND TOMORROW

Chapter 16, Text pages 464–493

INTERFACING WITH TECHNOLOGY

Technology will continue to increase exponentially. Energy will be used more efficiently. Transportation will be faster. Biotechnical systems will help cure diseases, feed billions of people, and extend the life span. Nuclear energy will be used more. We try to forecast the future. How can you become technologically literate to survive in this highly technological world?

SCIENCE: What have you learned in science this year that has helped you understand the technologies you have studied?

_____

_____

_____

MATH: What have you learned in math that has helped you work with technology activities?

_____

_____

_____

SOCIAL STUDIES: Explain how history relates to technology.

_____

_____

_____

LANGUAGE ARTS: What will happen if you cannot read or comprehend what you read? What will happen if you cannot write?

_____

_____

_____

ART: How has art influenced technology?

_____

_____

_____

IMPACTS FOR TODAY AND TOMORROW

Chapter 16, Text pages 464–493

TEST

TRUE/FALSE: Write the full word TRUE or FALSE in response to the following statements.

_____ 1. Outputs from a technological system can be undesirable and unexpected.

_____ 2. People are unable to determine whether technologies are good or bad by the way they use it.

_____ 3. Futurists are people who monitor existing technologies.

_____ 4. Ergonomics is a field that applies technology to the physical needs of human beings.

_____ 5. The technology that enables computers to make sounds like human voices is speech synthesis.

MATCHING: Place the letter of the correct response in the blank to the left of the term it matches.

_____ 6. Magnus LTA

A. Technique for forecasting the future.

_____ 7. Halophyte

B. Plant that can grow in salt water and is edible.

_____ 8. Maglev

C. Spherical airship capable of carrying over 50 tons.

_____ 9. Confluence

D. Vehicle that floats above its tracks because of strong magnets.

_____ 10. Delphi survey

E. The flowing together of technology systems.

IMPACTS FOR TODAY AND TOMORROW

Chapter 16, Text pages 464–493

ANSWER SHEET

/////////// KEY WORDS \\\\\\\\\\\

1. acid rain
2. speech synthesis
3. voice recognition
4. videoconference
5. futures wheel

6. Delphi survey
7. trends analysis
8. Personal rapid transit (PRT)
9. Smart houses
10. Intelligent buildings

/////////// REVIEW \\\\\\\\\\\

FILL IN THE BLANKS:

1. a. Expected and desirable
    b. Expected and undesirable
    c. Unexpected and desirable
    d. Unexpected and undesirable

2. a. Passing laws that promote or restrict a particular technology.
    b. Using a technology for peaceful or for destructive purposes.

3. a. Futures wheel
    b. Cross-impact analysis
    c. Delphi surveys
    d. Trend analysis

4. way they use it.

IDENTIFICATIONS:

5. Nuclear Regulatory Commission
6. High-definition television
7. Industrial Space Facility
8. Plain old telephone service systems
9. Lighter-than-air
10. Personal rapid transport

MATCHING:

11. E
12. D
13. H
14. A
15. B
16. G
17. F
18. C

TRUE/FALSE:

19. False
20. True
21. True
22. False
23. True
24. False
25. True

(Answers may vary and there may be additions to these suggested answers. Answers can depend upon the level of studies in the various disciplines for the group of students you are teaching.)

SCIENCE: Have students write this answer down after a class discussion on what has been learned in the school year or the previous school year that has helped them to better understand technology.

MATH: In math we learned how to calculate. This is important when figuring amounts of materials needed and the sizes required when working with scale models or other lab activities. We learned the significance of the word *exponential* in reference to the growth of technology. (Answer this question after a class discussion on what they learned in math.)

SOCIAL STUDIES: History tells the story of technology from the beginning of time to the present. History is also used to predict the future of technology.

LANGUAGE ARTS: If you cannot read or comprehend what you read, you will be unable to maintain and repair the products of technology. You will be unable to learn how to use the devices correctly. And, most importantly, you will be unable to read or fill out warranty cards and contracts for purchase. It is highly possible that others will know this and take advantage of your lack of ability to read and write.

ART: All technological devices must be designed. Manufacturers keep in mind the desire of the consumer to have a product that is pleasing in appearance. They also know that attractive advertisements will sell products. Our selection of a product is based somewhat on the appeal it has to us.

///////// **TEST** \\\\\\\\\\

TRUE/FALSE:

1. True
2. True
3. False
4. True
5. True

MULTIPLE CHOICE:

6. c
7. b
8. d
9. e
10. a

APPENDIX E
TECHNOLOGY
ACTIVITY GUIDELINES

NOW AND THEN

Contributed by Thomas Barrowman

Suggested Activity Length – 1 to 2 weeks

Activity Objective – This activity will help students to understand how technology influences our lifestyle today and how technological devices were used in the past.

Instructor Guidelines – Be aware of safety requirements and operating procedures when allowing a student to operate or demonstrate older tools or equipment.

Most technological devices improve the time factor in completion of a given job. The quality of the output is usually improved also.

Models of the tool or device could be constructed, as long as time spent is kept to a minimum and the center focus of the activity is the research and knowledge gained in accomplishing the activity.

Answers to Technology Connections

1. Complex technological systems develop from more simple technologies.
   Answer: Give students a few examples and let them supply others. Individual photographs to movies. Inboard motors to outboard motors. Manual control sewing machines to computer control. Radio to TV. Flintlock rifle to caplock to cartridge rifle. Single speed bicycle to ten-speed bicycle. Weather forecasting from observation to satellite images.

2. Was the original machine, tool, or process developed because of a want or a need? Did this want or need change with the modern version?
   Answer: Transportation land vehicles were developed in the early 1900s as a need. Time was an important factor in the transportation of people and goods. Today many land vehicles are developed as wants. Examples could be ATVs, ATCs, mountain bicycles, and racing cars. New models of cars come out each year and are societal wants.

3. Describe the technological developments that caused the tool or process you studied to be changed.
   Answer: Help the student pick significant technological developments that altered the tool or process. This may be hard for the student to find as the student researches the topic. Draw on your knowledge of the developments to help the students see the connections. You may want to do this with a time line.

4. How might the tool or process you've chosen change in the future?
   Answer: Let the students predict how the machine, tools or devices might change. Make connections between technologies and what might have to be developed before change could take place.

THE GREAT SPINOFF

Suggested Activity Length – 1 to 2 weeks

Activity Objective – In this activity students will construct a top and be involved in a competition to see whose top spins the longest. They will use the problem-solving process to construct the optimum top.

Instructor Guidelines – NOTE: Throughout this activity, students are to wear eye protection and follow all safety procedures.

Allow students a choice in the materials that they need to make their tops. Discuss the differences in types of wood, diameters of dowel rods and different string lengths and how these factors could affect the outcome of the top spinning.

Students might want to vary the factors in preliminary spin-offs before the actual competition. Have them keep notes on their findings.

When the competition is over, have students compare their tops to the winning top, and discover what they could have done to their top to improve its performance.

Answers to Technology Connections

1. List three changes you made to your top that helped increase the length of time that it remained spinning.
   Answer: Answers will vary for each student, but could include: changing length of string, the diameter of the top and its thickness, the length of the dowel, and the kind of point for it.

2. Compare your top to the one that remained spinning the longest in your class. What changes could you make to your top that might help it spin longer?
   Answer: See answer for Question #1.

3. Explain why a gyroscope is more than a toy.
   Answer: Gyroscopes are accurate navigational aids used to guide ships, submarines, and airplanes.

PLOP, PLOP, FIZZ, FIZZ

Suggested Activity Length – 1 to 2 weeks

Activity Objective – The purpose of this problem-solving activity is to allow the students to graphically demonstrate the use of systems and subsystems to solve a problem.

Instructor Guidelines – Gather all materials that are needed for this project. You may want to set certain size and weight restrictions on the project to give the students an even greater challenge.

Students could be divided into groups of two or three, or they may work on the project individually. Amount of time allotted for this project is an important deciding factor.

You may wish to judge students on their originality, neatness, and construction techniques as well as accuracy and the ability to complete the given task.

Answers to Technology Connections

1. Describe two problems that had to be solved in order to get your Alka-Seltzer delivery system to work.
   Answer: Two problems that had to be solved could include: achieving the proper amount of force needed to move the delivery system, and the construction of the delivery system so it will not fall over. Other factors to take into consideration would be weight, size, length to travel, pickup system, etc.

2. Describe each of the subsystems in your transport system.
   Answer: Each student will have different answers according to their individual transport systems.

3. Describe an alternative solution to the problem that you later considered inappropriate.
   Answer: Will vary with each student.

IT'S ABOUT TIME

Contributed by Alan Horowitz

Suggested Activity Length – 2 to 4 weeks

Activity Objective – This activity can be used as part of a unit on 'time'. Time is considered one of the seven basic resources of technology. In this activity, students will design and construct a clock using a battery-operated quartz movement. Students can also construct and assemble a variety of time measurement devices used throughout history.

Instructor Guidelines – Some of the devices you may wish to construct may be an hourglass (made from soda bottles, a cork with a hole drilled in the center, and some sand), a water clock (made from tin cans), a sundial, etc. *The Toy Shop* computer program (by Broderbund) has a neat little sundial activity you may wish to use.

Students should not use any tools or machinery until they have been instructed on their safe and proper operation. **Remind students that they should always wear eye protection when working in the technology lab.**

Each student should make a full-scale template from heavy paper or cardboard. The shape and size of the clock can be left up to the individual students. Make sure they leave enough room in their designs for the quartz action and numbers.

You may want students to make a detailed list of supplies, materials and resources they used to produce their clocks as they work through the activity. Students can compute the approximate cost of: supplies and materials, labor (minimum wage), machinery and tools, etc. for their clocks. Let students use current catalogs to look up the costs. If you have enough catalogs, this can be done as a group activity.

After the students have cut the shape of their clocks, have them use the belt sander, miter box or file to smooth and/or square the edges. Routing the edges is optional but usually makes the clock look nicer. Small orbital sanders save a lot of time and perspiration.

The clock parts (quartz action, numbers, hands, washers, nuts, sawtooth hanger, etc.) are available from a number of vendors, including:

Village Originals
24140 Detroit Road
Westlake, OH 44145-1570
(212) 836-2144

Klockit
P.O. Box 629
Lake Geneva, WI 53147
(800) 556-2548

If you order in a large quantity, you may be able to get a special lower price.

Answers to Technology Connections

1. The seven resources used in technological systems are people, information, materials, tools and machines, energy, capital, and time. Describe how you used each of the seven technological resources to design and build your clock.
   Answer: People: Labor (skilled and unskilled) and management.
   Information: Knowledge about safety, manufacturing sequence, direction from teacher, etc.
   Materials: Wood, metal, plastics, ceramics (quartz movement).
   Tools and Machines: Twist drill, router, jigsaw or band saw, drill press, and any other tools students might have used.
   Energy: Electricity to power machines, lights, etc.
   Capital: Money to purchase raw materials - wood, clock parts, finish, etc.
   Time: Learning the material (training) and production time.

2. Time is an important resource in the information age. The growing use of computers has made time even more important. Why?
   Answer: Some computers can process data in nanoseconds (billionths of a second). Computer access and on-line time has become very expensive.

MODEL ROCKET-POWERED SPACECRAFT

Contributed by A.R. Putnam

Suggested Activity Length – 3 to 4 weeks

Activity Objective – The purpose of this activity is to demonstrate how different subsystems, dependent upon feedback, work together to accomplish a task. The activity works best when students work independently and at their own speed. Complexity can easily be added to the assignment for more advanced students. If time permits, students should be encouraged to decorate their crafts when complete.

Instructor Guidelines – Model rocketry is safe, educational, and has high interest appeal with students. A few safety rules must, however, be followed:

1. Engines must only be ignited electrically and by remote control. Each package of engines contains instructions.
2. Rockets must never be fired indoors or in a congested area. A launch rod must be used, and no one should stand closer than 10 feet to the launch area.
3. Rockets should never be recovered from power lines or other dangerous places.
4. All vehicles should be tested for flight stability before being flown the first time.

Answers to Technology Connections

1. Which part of your craft acts as the low-speed guidance system? The high-speed guidance system? The recovery system?
   Answer: The launch lug (soda straw). The fins. The parachute.

2. Systems with feedback control are used by most industries today. What kind of feedback does the entertainment industry use? What about the advertising industry?
   Answer: Record and tape sales, attendance at concerts, attendance at theaters. Product sales.

3. Some outputs of some systems are undesirable. What are undesirable outputs from the air transportation industry?
   Answer: Noise and air pollution.

NOTE: Duplicate the following form as needed for the students to record feedback data on each flight.

MODEL SPACECRAFT FEEDBACK FORM

1. Lift off was straight and stable.

   _____ yes _____ no (See #7)

2. Acceleration was straight, smooth, and stable.

   _____ yes _____ no (See #8)

3. The craft glided to peak altitude after powered flight.

   _____ yes _____ no (See #9)

4. The ejection system activated shortly after apogee.

   _____ yes _____ no (See #10)

5. The recovery system ejected smoothly.

   _____ yes _____ no (See #11)

6. The recovery system functioned correctly.

   _____ yes _____ no (See #12)

7. Check alignment and attachment of straw. Check launch rod.

8. Check the fins: shape, alignment, thickness, connections. Add weight to the nose or increase the fin area.

9. Reduce vehicle weight or select a different engine (check the instructions that came with the engine).

10. Select a different engine.

11. Check the engine fit and nose cone fit. Use talcum powder to lubricate the recovery system. Pack the recovery system more loosely.

12. If burned, use more flame-proof wadding. If tangled, pack the recovery system better. Increase or decrease the size of the system as needed.

TUG O' WAR

Contributed by Thomas Barrowman

Suggested Activity Length – 3 to 4 weeks

Activity Objective – In this activity, students will learn about switching polarity with DC motors, pulley ratios, systems, and problem solving.

Instructor Guidelines – Students will find a greater challenge if the vehicles are placed near the start line and more in the direction shown by the arrows.

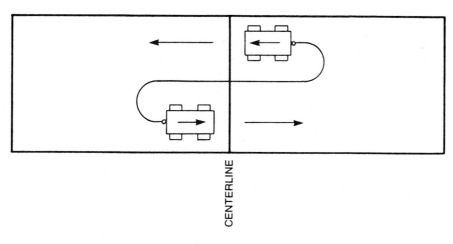

Vehicle set up

Rechargeable batteries can be used as long as you have enough spares for the competition.

The competition can be run on blacktop, rug, or waxed floor. Winners will be different on each surface.

On page 86 is the pictorial wiring diagram of a double-pole double-throw switch, single-pole switch, and a four-cell battery pack for each motor used (two motors pictured).

The DC motors will develop a lot of torque by using a 3/16" pulley on the motor and a 2" diameter pulley mounted on the axle. (You can obtain the motor from Edmund Scientific, 101 East Gloucester Pike, Barrington, NJ 08007.)

The final drive pulley can be part of a wood wheel that is turned on the lathe. The jig is made out of 1/4 – 20 machine bolt 4" long with two nuts and washers used to hold the wood in place. Center drill one end for a live center to be used in the tailstock of the metal lathe.

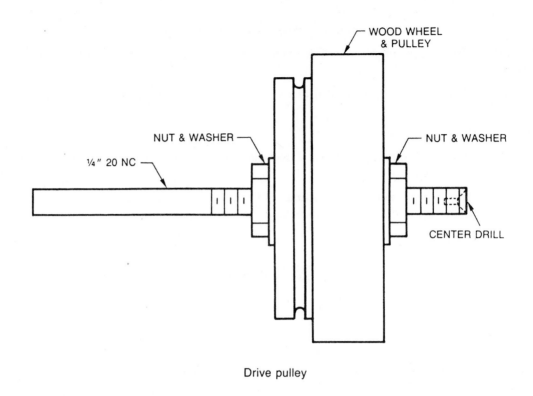

Drive pulley

Answers to Technology Connections

1. Problem solving begins by identifying the problem. In this activity, what was the problem?
   Answer: The problem was to design a vehicle with the most traction and power.

2. What goals did you set? What specifications did you have?
   Answer: The stated goal was to pull an opponent's vehicle over the center line. For specifications, the student should list overall dimensions, types of wheels, chassis material, and how many pulleys to be used.

3. What alternative solutions did you come up with?
   Answer: Alternative solutions could be to lighten the vehicle, use a wider wheel base, or use a mousetrap to snap the string.

4. Why did you choose your solution?
   Answer: The best solution will be the one that does best in trial runs.

5. What feedback did you receive? How would you change your solution for next time?
   Answer: The student should list test results from trial runs. The student should examine the feedback data and make appropriate adjustments.

6. Lubricants help reduce friction. Are there areas on your vehicle where you could apply lubricating compounds? What type of lubricant would you use?
   Answer: Lubricant can be applied to pulleys and wheels. Graphite, teflon, silicon, and WD-40 are lubricants that could be used.

7. Increasing the weight of the vehicle may give you an advantage. How will this help? What vehicles are designed with a heavy chassis?
   Answer: Increased weight over the drive wheels would give better traction. Bulldozers, tow trucks, bucket loaders, and tractors are some vehicles with heavy chassis.

SPACE STATION

Contributed by Margaret Rutherford

Suggested Activity Length – 4 to 5 weeks

Activity Objective – This activity gives students an opportunity to learn about space exploration and the impact it has on manufacturing, construction, communications, and transportation technologies. The student will develop an appreciation of the value of NASA's research and its development of new technologies.

Instructor Guidelines – Proper preparation is the key to success in teaching the Space Station activity. A prerequisite for this activity is the development of an environment conducive to stimulating creative thought processes. The physical environment, materials, literature, and equipment need to be in place. Below is a list of suggestions to enable you to achieve a successful unit of instruction.

1. From NASA obtain posters, pictures of astronauts, shuttles, and other related materials. Arrange these on the walls, bulletin boards, windows, entry to the lab, and on the door.
2. Prepare overhead transparencies covering the following:
   a. Instructional procedure you will follow
   b. Objectives to be covered
   c. Modules and components of the Space Station
   d. Uses of the Space Station
   e. Sample proposal for students to use as a guide
   f. Safety to be followed in using all tools, equipment, and materials
   g. Laboratory management
3. Order classroom sets of NASA material concerning the Space Station or make enough copies so each team has a set of literature in addition to the text materials.
4. Order films or videotapes that relate directly to the Space Station.
5. Collect examples of background music for the videotape presentation of each team's Space Station.
6. Prepare yourself intellectually with all the information on the technologies involved with the Space Station.
7. Present the unit with enthusiasm! Greet each student each day to make each feel he or she is an important member of the space station teams.

This activity is divided into two parts. Designing and constructing the Space Station is the first part, and the presentation of the Space Station on videotape is the second part. This is a sample format for this activity:

1. Cover the objectives or justifications for teaching the activity. (The Space Station involves the study of communications, energy, power, transportation, manufacturing, and construction technologies.) Using an overhead transparency of the Space Station, show how the Space Station with its dual keel configuration is similar to a football field in shape and size. Using an overlay of a football field, show positions of the modules and components of the Space Station.
2. By means of lecture, students reading, films, videotapes, or current events, present information on the current proposed Space Station.
3. Check the students for understanding of materials covered. Have discussion, verbally spot check students, or use strategies to ascertain if the students are assimilating information being presented.
4. Divide each class into teams of 4-6 students per team. Have all members of the class draw up designs for a space station. Then follow the procedures as outlined in the PROCEDURE section of the text.
5. Team members will be working simultaneously on the construction, proposal and videotape presentation. As needed, provide additional instruction on these. Instruction can be one-to-one or addressed to the entire class.

6. Have videotaping and recording equipment and computers available. You might have a guest speaker from a television station come into the class and work with the students on the presentations of the space stations on videotape.

7. After all the space stations are complete and videotape presentations are made, have an open house for parents, the community, school board members, and administrators. During open house, run the videotape of all the presentations.

8. Invite the news media to come visit the classes at any time during the construction phase. Keep abreast on current news on the Space Station. The news media are more prone to do a news story in an educational setting when it relates directly to real world situations. Have ready materials and a press release relating the activity. Be ready to answer questions concerning the Space Station and its relationship to the curriculum and to justify the teaching of such an activity.

NASA RESOURCES:

MAGAZINES AND PAMPHLETS: NASA Spinoff 1985, NASA Spinoff 1986, PMS - 008 Space Station, EP - 211 Space Station, EP -251 The Tracking and Data Relay Satellite System, Social Sciences and Space Exploration, Living Learning Working in Space, NASA Film/Video Catalog

FILMS AND VIDEOTAPES: JSC 880 U.S. Space Station Assembly Scenario, CMP 111 Space Station: The Next Logical Step, CMP 094 NASA Space Station: Preliminary Conceptual Designs, VCL 1180 Model, Double Keel Space Station, Resource (There are numerous films and videos of Skylab, the First United States Space Station. Titles may be obtained from the field catalog.)

For publications and address of the nearest NASA facility which will send publications to you, write:

Superintendent of Documents
U.S. Govt. Printing Office
Washington, DC 20402

Answers to Review Questions

1. What are the functions of the following modules and components of the space station?
   a. Habitat - living quarters
   b. Laboratories - scientific research
   c. Servicing Facility - repair and service of spacecraft and satellites
   d. Solar Dynamic Power Unit - converts sunlight into electrical energy
   e. Vertical Keels - the parallel beams that provide the support of the upper, lower, and middle beams to which are attached the modules, photovoltaic arrays, solar dynamic units, and antenna
   f. Photovoltaic Arrays - collect the sun's energy and convert it into electrical energy
   g. Remote Manipulator - assemble and maintain the Space Station
   h. Radiator - dissipates heat
   i. TDRSS Antenna - (Tracking and Data Relay Satellite System) makes voice exchange and record data flow possible

2. List eight possible uses of the space station that would justify its existence.
   a. Scientific research
   b. Develop space-based communication systems
   c. Observatory to study the Earth and sky
   d. Repair and service satellites
   e. Manufacture of metallic super alloys for construction, pure glass for laser and optical uses, super-pure pharmaceutical chemicals, and superior crystals for electronic systems
   f. Serve as construction site to assemble structures too large to be carried by a space shuttle
   g. Serve as a base for vehicles that will send and retrieve payloads to and from a higher orbit
   h. Serve as a living quarters

3. Which foreign countries are involved with the United States in the construction of the space station? What are the contributions of each?
   a. Japan is designing a research and development laboratory.
   b. The European Space Agency is designing a laboratory and unmanned platform.
   c. Canada is designing a Mobile Servicing Center equipped with a remote manipulator arm.

SPACE STATION

AND TEACHER HELPS

SPACE STATION OBJECTIVES

1. IDENTIFY MODULES AND COMPONENTS.
2. DESCRIBE FUNCTIONS OF MODULES AND COMPONENTS.
3. LIST USES OF THE SPACE STATION.
4. DESIGN AND CONSTRUCT A MODEL OF A SPACE STATION.
5. DISCUSS INTERNATIONAL CONTRIBUTIONS TO THE SPACE STATION.

 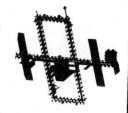

SPACE STATION

1. VERTICAL KEELS AND HORIZONTAL BOOMS

2. LABORATORIES

3. HABITAT

4. SOLAR DYNAMIC POWER UNIT

5. SERVICING FACILITY

6. TDRSS (TRACKING & DATA RELAY SATELLITE SYSTEM) ANTENNA

7. REMOTE MANIPULATOR ARM

8. RADIATOR

9. PHOTOVOLTAIC ARRAYS

FUNCTIONS OF THE SPACE STATION

1. LEADERSHIP IN SPACE

2. CONSTRUCTION SITE TO ASSEMBLE STRUCTURES TOO BIG FOR SHUTTLE DELIVERY

3. REPAIR OF SATELLITES

4. OBSERVATORY TO STUDY THE EARTH AND SKY

5. SPACE-BASED SYSTEM OF COMMUNICATIONS

6. SCIENTIFIC RESEARCH

7. MANUFACTURING PLANT FOR PRODUCTION OF:

Metallic super alloys

Pure glass for lasers

Pharmaceutical chemicals

Superior crystals

8. BASE FOR LAUNCH AND RETRIEVAL OF PAYLOADS

PROPOSAL FOR THE SPACE STATION

1. FRONT COVER WITH NAME OF THE SPACE STATION

2. LIST SUB-CONTRACTORS & TEAM MEMBERS

3. SKETCH OF SPACE STATION

4. LIST OF MATERIALS, TOOLS, AND EQUIPMENT

5. PROCEDURE FOLLOWED IN CONSTRUCTION OF THE SPACE STATION

6. LIST OF COMPONENTS AND MODULES OF THE SPACE STATION AND FUNCTIONS OF EACH

VIDEOTAPE SPACE STATION

1. WRITE UP SCRIPT
    A. Name of team
    B. Members of team
    C. Procedure followed
    D. Parts of space station
    E. Functions of station
2. CHOOSE MUSIC FOR VIDEO
3. BUILD SETTING & PROPS
4. REHEARSE PRESENTATION
5. VIDEOTAPE PRESENTATION

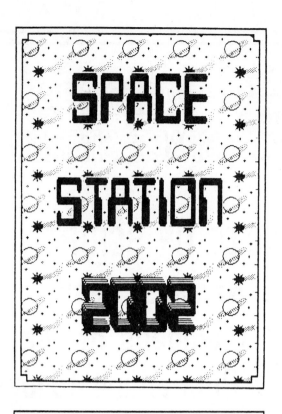

SPACE STATION 2002

HOWELL INTERMEDIATE
TECHNOLOGY I CLASSES
INVITE YOU TO AN EXHIBIT

SPACE STATION 2002

AT HOWELL INTERMEDIATE
ROOM 306
MONTH DAY, YEAR
FROM 8:00a.m. - 5:00p.m.

 SPACE STATION

OF SPACE STATION _____

NAME OF SPACE STATION _____

READ EACH INDICATOR. PLEASE CIRCLE THE NUMBER WHICH
INDICATES THE LEVEL OF ACHIEVEMENT OF EACH ITEM. FEEL FREE
TO MAKE COMMENTS AT THE BOTTOM OF THIS PAGE.

```
5 ..... CLEARLY OUTSTANDING
4 ..... OUTSTANDING
3 ..... ACCEPTABLE
2 ..... UNACCEPTABLE
1 ..... MISSING
```
                                                            POINTS

1. OVERALL APPEARANCE. 1   2   3   4   5   ____

2. HABITATIONAL QUALITIES. . . . 1   2   3   4   5   ____
    (COULD ONE EXIST IN THE
    SPACE STATION WITH THE
    MODULES AND DEVICES AS
    EXHIBITED?)

3. PRACTICALITY OF DESIGN . . . 1   2   3   4   5   ____
    COULD THIS DESIGN BE
    ASSEMBLED IN SPACE? WOULD
    THE ARRANGEMENT OF MODULES
    BE PRACTICAL?

4. CONSTRUCTION 1   2   3   4   5   ____
    IS THE WORKMANSHIP NEAT?
    ARE PARTS WELL ATTACHED?

5. FINISH 1   2   3   4   5   ____
    IS THE PAINT JOB NEAT?
    ARE MODULES AND PARTS OF
    THE STATION NEATLY LABELED?

6. LEVEL OF DIFFICULTY 1   2   3   4   5   ____
    HOW DOES THIS COMPARE TO
    ALL OF THE OTHER SPACE
    STATIONS IN THE COMPETITION?   _____

                                            TOTAL   ____

COMMENTS_____

        _____

381

SPACE STATION CROSSWORD

ACROSS:

6. TYPE PANEL THAT DISSIPATES THE HEAT OF THE SPACE STATION
7. COUNTRY THAT IS DEVELOPING PRESSURIZED LAB MODULE FOR THE SPACE STATION
8. INITIALS FOR THE EUROPEAN SPACE AGENCY
11. PRINCIPAL MEANS OF TRANSPORTATION TO AND FROM THE SPACE STATION
12. A LAB, HABITAT, ASSEMBLY FACILITY, OR STORAGE DEPOT.
13. TYPE OF PANEL THAT POWERS THE SPACE STATION

DOWN:

1. MODULE ON THE SPACE STATION WHERE PEOPLE LIVE
2. NATIONAL AERONAUTICAL SPACE ADMINISTRATION
3. SYSTEM THAT CAUSES MOVEMENT THROUGH SPACE
4. MODULE OF THE SPACE STATION WHERE EXPERIMENTS ARE CONDUCTED
5. A VEHICLE RUN BY COMPUTERS. PEOPLE ARE NOT PRESENT IN THE VEHICLE.
9. A ROBOTIC ARM OF THE SPACE SHUTTLE MADE BY CANADA
10. COUNTRY ABOVE THE U.S. THAT IS PARTICIPATING IN THE CURRENT SPACE STATION
11. A STOPPING POINT IN SPACE THAT WILL HELP EXPLORATION

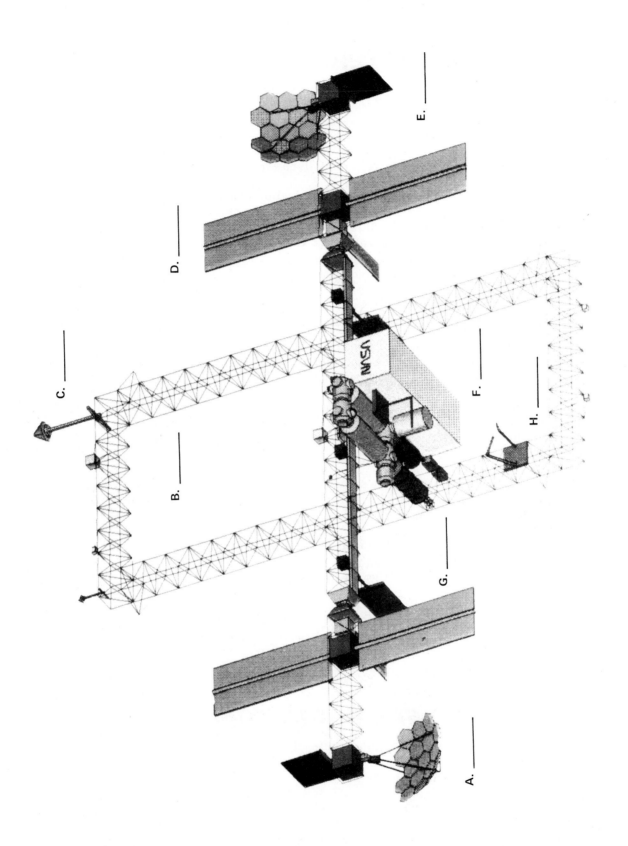

A.

B.

C.

D.

E.

F.

G.

H.

383

COMPUTERS IN INDUSTRY

Contributed by Thomas Barrowman

Suggested Activity Length – 2 to 3 weeks

Activity Objective – In this activity, students will use the computer as a tool to test production lines and vehicle designs. Drafting with a computer will be explored.

Instructor Guidelines – Most labs have only a few computers available for the students to use. It is recommended that these activities be introduced as soon as possible in the course.
   "The Factory" and "Car Builder" will introduce the students to industrial uses of the computer.
   Incorporate the design and testing features of "Car Builder" by modeling the designed car. A clay or paper mache' model can be constructed and the design tested in the wind tunnel. The student will then have a real experience in CAD development and actually carry the computer simulation to the developed model.

Answers to Technology Connections

1. After using "Car Builder," model the body you designed. Test it in the lab wind tunnel.
   Answer: The model should be the exact same scale as the printout on the computer. Comparisons can then be made with the air flow patterns over the top of the car.

2. Name and properly hook up all the computer accessories that you used for the above programs.
   Answer: Supply the students with a pictorial drawing of each component in the computer system. They can then draw in the wires needed (computer, joy stick, printer, mouse, power supplies).

3. What other communication systems are used in combination with the computer?
   Answer: Telephone lines, satellite transmission, and graphic printing machines are used in combination with the computer.

"The Factory" is available from:
Sunburst Communications
39 Washington Avenue
Pleasantville, NY 10570

"Car Builder" is available from:
Optimum Resource, Inc.
Middletown, CT 06457

LOW-POWER RADIO TRANSMITTER

Contributed by Alan Horowitz

Suggested Activity Length – 4 to 6 weeks

Activity Objective – In this activity students will construct a small radio transmitter in order to become familiar with electronic communications, components, printed circuits, and construction techniques. This radio transmitter will be able to send a message through a standard A.M. broadcast band radio.

Instructor Guidelines – You may wish to duplicate the full-size printed circuit board layout shown in the text. Give a copy to each student involved in this activity. Tell your students not to write in the textbook!
   This is a *very low-power* radio transmitter and the range is short. If a longer antenna is used, the range will increase.

COMPONENT LIST

T1	2N3906 PNP Transistor (Radio Shack #276-1604)
R1	130 Kohm resistor (Kelvin #1/2 watt 130,000)
C3 & C4	.022 ufd capacitors (Kelvin #Mini Disc .022)
C2	.0047 ufd capacitor (Radio Shack #272-130)
C1	100 pfd (uufd) capacitor (Kelvin #Disc 100)
L1	Adjustable Tapped Loopstick antenna coil for broadcast band (Kelvin #185-76)
L2	10 or 15 turns of #30 enameled wire around L1
B1	9-volt alkaline transistor battery with 9-volt battery snaps
M1	1000 ohm magnetic-type earphone or microphone with 1/8" phone plug
S1	Miniature spst toggle switch
ANT	30" Telescoping antenna (Radio Shack #270-1401)

1/8" Miniature phone jack - open circuit (Radio Shack #274-297)

The component lead positions and lengths can affect the transmitting power and frequency. Final tuning is very touchy.

Different PNP amplifier transistors can be substituted for the 2N3906 transistor. You may wish to experiment with other components.

Instead of using a printed circuit board, the transmitter can be constructed on perfboard and point-to-point wiring used.

There are many other low-power A.M. wireless transmitter circuits available in magazines and electronic experimenter books. Modern Electronics magazine ran a multi-part series on "Low-Power Radio Transmitters" by Forrest M. Mims in some of the 1986-87 issues. *The ARRL Handbook for the Radio Amateur* has detailed information about the design and construction of low-power transmitters. It is available from:

American Radio Relay League
225 Main Street
Newington, CT 06111

The circuit used in this activity is one of the simplest possible. For higher power transmitters or circuits that are more versatile and stable, I would suggest constructing a multi-transistor transmitter. Wireless transmitter kits are available from a number of resources, including:

Kelvin Electronics Inc.
P.O.Box #8
1900 New Highway
Farmingdale, NY 11735
(800) 645-9212
(516) 249-4646 (in New York)

and

Pitsco
P.O. Box 1328
Pittsburg, KS 66762
(800) 835-0686

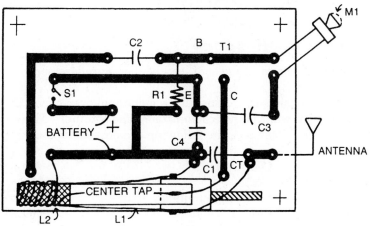

X-ray view—from the foil side of the circuit board

Answers to Technology Connections

1. Electronic circuits are made up of components. Each component has a specific function in the circuit. What is the function of the microphone in the transmitter? The battery?
   Answer: The microphone converts sound vibrations into electrical impulses. The battery supplies electricity to other components in the circuit.

2. The use of electronics has revolutionized all aspects of technology. What would your life be like without electricity? What things would you miss the most?
   Answer: No radios, record players, electric lights, T.V., power appliances, automobiles, computers, VCRs, etc.

3. What is it called when many components of a circuit are miniaturized and produced on one piece of semiconductor material?
   Answer: An integrated circuit.

INTERNATIONAL LANGUAGE

Contributed by Thomas Barrowman

Suggested Activity Length – 3 to 4 weeks

Activity Objective – Students will learn how to use graphic communication devices in their development of a pictorial direction booklet.

Instructor Guidelines – Language differences have long been a possible barrier to effective communication. A few years ago, directions for such things as ski equipment and other internationally sold merchandise were printed simultaneously in German, Japanese, Spanish, Italian, French, and English. The world trade market has expanded to include so many countries that increased printing costs have resulted. Pictorial directions have emerged as a cost-effective alternative to printing directions in several languages.

An excellent example of a set of pictorial directions can be found in use with the Mistral wind boards. (See your local dealer for a copy.)

Many cameras have picture directions. Wordless Workshop in Popular Mechanics magazine has been successful for years. Airlines use pictures for explaining how to use safety equipment and for escape routes. Restaurants post procedures to aid choking victims in pictorial form.

One of the first activities you should involve the students with is interpreting a set of directions before the students produce their own directions. Have the students develop a set of directions and utilize the graphic communication equipment and supplies you have available to print the booklet.

The students can contribute commercial examples of pictorial directions that you can keep in your "idea" file.

Answers to Technology Connections

1. Communication includes having a message sent, received, and understood.

2. The communication process consists of a transmitter, a channel over which the message travels, and a receiver. In this activity, what is the transmitter? The channel? The receiver?
   Answer: Depending on the process used, the following are possible answers:

Transmitter	Channel	Receiver
pen	ink	paper
computer printer	toner, ink or ribbon	paper
light focused by camera	film	snapshot
printing press	ink	paper

3. What purpose (inform, persuade, educate, or entertain) does your communication system accomplish?

   Answer: The purpose of your communication system is to inform.

4. How did you use each of the seven technological resources to create your communication system?

   Answer: People - yourself, the librarian, your parents, other students.

                  Information - library, books, direction booklets, product manuals.

                  Materials - paper, film, ink.

                  Tools and machines - computer, drafting equipment, copying machines.

                  Energy - electrical energy for machines, human energy.

                  Capital - money spent for materials used.

                  Time - ten hours.

TEAM PICTURE

Contributed by Alan Horowitz

Suggested Activity Length – 1 to 2 weeks

Activity Objective – In this activity, students are asked to design and produce a stencil or iron-on transfer using a computer and graphics software. This stencil or iron-on will be used to personalize a T-shirt or sweatshirt in order to graphically communicate a message. A contest for the best design can be held.

Instructor Guidelines – If you are not familiar with the computer, software and printer you will be using with your students, take some time to read the manufacturer's documentation and go through the start-up, composition, and printing procedures in advance. If you have any problems, ask another teacher or one of your students for help. You'll be amazed how much some of your students in your classes know about computers.

Some of the software programs you can use for this activity include: PrintShop, Dazzle Draw, Mousepaint, MacPaint, FullPaint, Zoom Graphics, Special Effects, Kaola Draw, Triple Dump, Blazing Paddles, etc.

While some students are quite computer literate, it will probably be necessary to give some other students a basic introduction to the computer, i.e., proper handling of the disks, start-up procedures, loading and running programs, printing material, etc. Caution students about bringing a floppy disk near a magnetic field, especially if you are using a graphics tablet. A graphics tablet can become an electromagnet when it is activated.

Make sure your computer disks are write-protected and, if possible, use back-up disks with students instead of the originals.

Walk students through the software program one step at a time. A succinct handout of the step-by-step procedure to be used will be <u>very</u> helpful. The handout will save a lot of the "What do I do next?" questions.

A number of colors for the Underware^(TM) direct iron-on transfer ribbons are available. Underware color pens are also available to add color to the transfers. These items are available from:

Diversions, Inc.
505 W. Olive Ave. #520
Sunnyvale, CA 94086
(408) 245-7475

An assortment of thermal screens for the thermofax machines, frames, squeegees, and textile paints are available from:

Welsh Products, Inc.
1201 E. 5th Street
P.O. Box 845
Benica, CA 94510
(707) 745-3252

1. Computer-generated graphics, including animation, are important in today's communications. How is this technology used by the entertainment industry? What about the advertising industry?
Answer: Many motion pictures make use of computer-generated graphics. Video games and T.V. also make use of computer-generated graphics for entertainment and advertising.

2. Scientists and engineers use computer-generated graphics and computer aided drawing (CAD) to help plan and test new automobile and aircraft designs. How does this help keep the design costs down?
Answer: Using computer simulation is much cheaper than constructing and testing the actual vehicle. Design changes can be made on the computer screen before committing a lot of money to prototype production costs.

3. Computer graphics and text are often combined in desktop publishing. What is desktop publishing? Do you think it will play an important part in the communications field?
Answer: With desktop publishing, professional-looking documents can be produced with a computer and laser printer. Companies may save a substantial amount of printing/copying costs by publishing documents in-house.

LEARNING ABOUT CAMERAS AND FILM DEVELOPMENT

Suggested Activity Length – 1 to 2 weeks

Activity Objective – In this activity, students will create charts and graphs in order to compare features on different cameras and be able to use that information to purchase or use the most desirable camera for their needs.

Instructor Guidelines – The students should be put into groups of 2 to 4 to reduce the number of brochures needed.

Have the students chart or graph their comparisons in a number of different ways (i.e., bar graph, pie chart, etc.). The charts and graphs could then be posted in the classroom so the classes can compare the work they did with that of other classes.

Be sure that all directions are followed when film is loaded, exposed and developed. The students should again work together in groups.

Answers to Technology Connections

1. Advances in optics have reduced size and weight of the cameras, increased the ease of focusing and the brightness of the viewing screen, and allowed auto focus and zoom lens production.

2. How has the creation of new plastics and other man-made materials affected the size, weight, and strength of more modern cameras?
Answer: Cameras have become smaller in size, lighter in weight, and more durable (able to take being dropped, though this is not advised). Cameras now come in all sizes, shapes, and colors, and are made from many different materials. Cameras have even become disposable.

3. Do you think that the use of computers, both for design and within the camera itself, has affected the types of cameras produced today?
Answer: Yes, the use of computers has not only helped in the design (smaller, easier to use), but also in the quality of picture that can be taken. One does not have to be a professional to take professional-quality pictures.

PHOTOGRAMS

Contributed by Fred Posthuma

Suggested Activity Length – Depending on the number of students in the class and the equipment, this activity should take from 2 to 4 class periods to complete.

Activity Objective – After completion of this activity, the student should understand the basic process of photography as it relates to graphic communications.

Instructor Guidelines

1. Have the chemicals mixed prior to the students' activity.

2. Remind <u>all</u> students that the photographic paper is light sensitive, except with the safelight. If a light is turned on before set-up is ready, the sheet(s) of paper will be exposed.

3. Have the photographic paper in small plastic (light proof) bags with one or two sheets per bag. This way, if lights are accidentally turned on, the entire package of paper is not ruined.

4. Use an RC (resin coated) paper. This means it can air dry and <u>does not</u> require a special dryer.

5. If using an enlarger, explain its usage prior to the activity.

6. Use a darkroom or some other room that is light proof. A sink with running water is also useful.

7. Use an OC safelight lens. This is typical for black-and-white paper development.

8. Arrange and label trays and have them organized in order of usage.

9. Let the students be creative!

10. This would be a good lead-in activity for making a photographic print from a negative.

Answers to Technology Connections

1. What are some commercial and industrial uses of photography?
   Answer: Commercial applications include architecture, real estate, publishing, commercial advertising, professional photo studios, etc. Industrial applications include inventory, product development, advertising, etc.

2. How has current technology altered photography? Hint: What about camera types, styles, and uses? Instant photography?
   Answer: Technology has altered the style and size of cameras over the years. Also, film processing has changed with the development of better films, as well as the introduction of "instant" photography. Microprocessors in cameras provide automatic operation.

3. How do you think large photographic processing companies process your film and prints?
   Answer: Large commercial development and printing plants typically are highly automated with strict processing standards adhered to. Only special situations require hand processing.

4. Why is photography a unique communication process as compared to other methods of communication?
   Answer: Photography has the ability to "freeze" a moment of time that will never happen again. It is a highly visual and unique mode of communication that records moments of time.

5. Why did some objects cause the paper, when processed, to appear gray, not really black or white?
   Answer: If the object used in the photogram is entirely flat and not translucent or transparent, the object will appear white due to no light hitting that surface. If the object is not entirely flat or translucent, it will appear somewhat gray.

MORSE CODE COMMUNICATIONS

Contributed by Alan Horowitz

Suggested Activity Length – 2 to 4 weeks

Activity Objective – In this activity, students will design and construct a traditional telegraphy device to demonstrate one method of electronic communication.

Instructor Guidelines – If necessary, lantern batteries can be used for a power source in place of the 6–12 volt DC power supply.

The sounder coil can be wound from almost any thin-gauge insulated wire. Plan on using a lot of wire with your students.

Lamp cord can be used to connect the telegraph key to the sounder. Run the lamp cord wire from one room to another so students can only communicate with each other using the telegraph unit.

You may wish to make a chart and/or handout of the international Morse code. Students can refer to the chart during this activity.

If the drill press is used to make the holes in the wood bases, it will save time if a jig is clamped to the drill press table.

This activity can be done with individual students or as a mass production project. A number of "assembly lines" can be set up and run simultaneously, each line making different parts for the key or sounder units. A final assembly line composed of the entire class can be used to complete the telegraph units.

If you hold the sounder unit next to an A.M. radio receiver while code is being sent, you can get an idea of how an old time spark-gap transmitter sounded. The spark made by the buzzing action of the sounder is sending out very low-power radio waves and this, in turn, is picked up by the A.M. radio receiver. Lightning during an electrical storm makes similar "crackling" sounds in a radio.

A simplified version of this activity can be made by replacing the sounder with a 12-volt lamp wired in series with the key and battery. The light will go on when the telegraph key is pressed. A 12-volt commercial buzzer can be used in place of the sounder.

If students are interested in learning more about Amateur Radio (ham) communications, free material can be obtained from the:

American Radio Relay League
225 Main Street
Newington, CT 06111

Answers to Technology Connections

1. The communication process consists of a transmitter, a channel, and a receiver. In telegraphy, the Morse code key is the transmitter and the sounder is the receiver. What part of the telegraph circuit represents the channel, or route, the message takes?
   Answer: The wire connecting the transmitter and sounder.

2. When telegraph lines are run long distances, relay stations are needed. Why?
   Answer: The resistance in a wire carrying electricity can cause a substantial voltage drop. A relay station is used to amplify or step-up the voltage from this loss.

BEAM THAT SIGNAL

Contributed by Thomas Barrowman

Suggested Activity Length – 3 days

Activity Objective – In this activity, the student will learn geometric construction, laws of reflection, laser safety, and an understanding of laser communications.

Instructor Guidelines – Heathkit has a well-built laser in kit or assembled form. A laser receiver can also be purchased from the Heathkit Company (800-253-0570). The Laser Technology course from Heathkit will give all the background information needed plus many other experiments the students can perform. You can extend this laser activity by having students do the following simulated mountain communication project. Light-protective glasses should always be worn when working with lasers.

Mountain Communication

Acrylic or glass mirrors 4" x 4" can be mounted with contact cement on 3/4" plywood. Drill a 1/2" hole 1-1/2" deep in the edge of the plywood and near the back. Cut a 1/2" steel rod 1-1/2" long to fit the drilled hole. This will balance the mirrors by adding weight to the back of the plywood.

The drawing below is the layout that the students will construct on the white board before the laser is turned on.

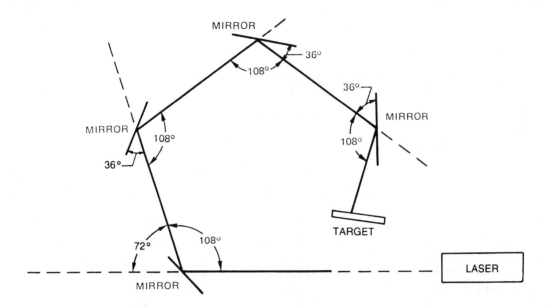

Laws of Reflection: In every case of reflection (1) the angle of incidence (i) is equal to the angle of reflection (r), and (2) the incident ray, the normal, and the reflected ray lie in the same plane.

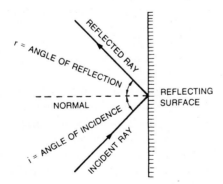

Angle of incidence equals angle of reflection

Answers to Technology Connections

1. Hook a piece of fiber-optic cable between the laser and the receiver. Plug the tape recorder into the input jack of the laser. Again play the tape recorder. This will demonstrate fiber optic communication.

2. Try the same experiment over a greater distance, such as 30 feet. Omit finding the angles. Keep all the equipment on the floor for safety. The laser beam must not hit your eyes.

3. How does a compact disc make use of a laser to record and play back information?
   Answer: A laser beam is used. The signal is coded onto a flat disc in the form of microscopic pits. The information contained in the pits is read off the disc by a laser. The pits, which are metallized, reflect the light to different degrees, depending on their depth. This reflected light is picked up by the photodiodes, which produce an electrical signal corresponding to the variations.

4. Fiber-optic cable is becoming increasingly common for what type of communication? How is the laser used in this type of communication?
   Answer: Fiber-optic cable is used for telephone and data networks. Information is put onto the laser light wave by varying the voltage driving the laser. A photodiode is used at the receiver to convert the light into electrical energy.

THE MISSING LINK

Contributed by Alan Horowitz

Suggested Activity Length – 2 to 4 days

Activity Objective – In this activity, students will use a computer, modem, and phone line to transmit and receive data. By participating in this activity, students will be exposed to an advanced form of electronic telecommunications.

Instructor Guidelines – If you are not very familiar with computer telecommunications (this simply means one computer talking to another!), sit back in a quiet location and familiarize yourself with the literature that came with your computer and modem. If you can find a teacher or one of your students who has a computer and a modem at home, he or she may be very helpful.

Whatever you do, do not get bogged down with the terminology used in computer telecommunications. It is not absolutely necessary to understand everything before getting started. You will almost certainly pick up the lingo as you go along. Your computer is connected to another computer using a *modem* (*MOD*ulator + *DEM*odulator = MODEM) and telephone lines.

Since each brand of computer, modem, and telecommunications software package is different in many ways, there are a substantial number of variables that can affect the exact procedure you will follow when interfacing and using your telecommunications system.

You must become familiar with the telecommunications software before you attempt to call another computer. (Here is where another teacher or one of your students can be of the most help.) Read the directions carefully.

Watch for key words like: baud rate (transmission and reception speed of the modem), dialing a number (touch-tone or pulse dialing), uploading (sending a message), downloading (receiving a message), full-duplex (usually for connecting to a mainframe computer), half-duplex (usually for calling another microcomputer), disconnecting or hanging up, etc.

You may wish to discuss the importance of computer telecommunications with your students. Resources like Dow Jones, Compuserve, The Source, TechNet, GEnie, Electronic Mail, 24-hour banking, and Grolier's Encyclopedia can be discussed. The rapid exchange of data is a very important part of our technological society.

Answers to Technology Connections

1. The communication process consists of a transmitter, a channel, and a receiver. In data communications, a computer with a modem is the sender, and another computer with a modem is the receiver. What part of the system represents the channel, or route, the message takes?
   Answer: The phone line.

2. A number of computers can be connected by telephone lines to form a network. What device is needed to connect a computer to a telephone line?
Answer: A modem.

3. Why is it important for all computer manufacturers to use a standard code (such as ASCII) when designing new computers? What would happen if every computer used a different code?
Answer: If computer codes were not standardized, each computer would only be able to successfully communicate with computers or peripheral devices (modems, printers, disk drives, etc.) of the same brand.

COMMUNICATING A MESSAGE

Contributed by Ethan Lipton

Suggested Time Frame – This activity, made up of several parts, may be used in several manners based upon the depth required in your situation and/or the time/resources available. These include: (1) each student may complete all of the parts of the activity individually; (2) the class may be divided into groups (size to be determined by the teacher), each group completing the entire activity; or (3) the class may be divided into three groups (not necessarily of equal size), each group completing one of the three components of the activity.

While time requirements will vary based upon your specific situation (resources, class/group sizes, depth of project objectives), the following suggested activity times are provided as guides based upon the three possible methods of application:

(1) individual completion of the entire activity: 30–40 hours
(2) a number of groups, each group completing the entire project: 20–30 hours
(3) class divided into three groups, each completing one component of the activity: 10–15 hours

Activity Objective – Students will utilize current electronic and graphic communication technology to produce a newsletter, audio presentation, and video presentation. These activities will involve them in the activities (ideas, concept, development, production) used in the production of newspapers, radio broadcast, and television.

Instructor Guidelines

GENERAL

How you use this activity will depend upon many factors including those presented in the discussion of a suggested time frame. Carefully consider all appropriate factors and tailor the experience to meet the instructional needs of your students and program.

It is very important that you "do" each part of this activity before using it in your class. This will help you develop any additional instructional components and may provide sample materials you may wish to use for your class. It will also facilitate adaptation of the procedures to the specifics of your environment.

It would be helpful to the students if you would prepare a working schedule for this activity. Tied into the importance of scheduling in the "real world," this will provide direction for the students and help them as a yardstick of their progress. If more than one student will be involved in the activity, discuss the importance of each team member's responsibilities (jobs).

If any of the processes and/or procedures of this activity are new to you, you may wish to consult texts and resources dedicated to those specific areas.

NEWSLETTER

- Provide students with enough information about, and samples of, rough layouts to maximize student understanding and aid in their development and use.
- Stress the importance of decision-making when producing the rough layout and following those decisions towards completion of the project.

- Make sure careful attention is placed on the editing and proofing stages of production.
- Emphasize the importance of proper positioning at the image assembly state (manual paste-up or desktop publishing). All lines must be straight and accurate.
- Discuss how these procedures are related to publishing (newspapers, magazines, books, etc.).

AUDIO TAPE

- Discuss the important aspects of script writing.
- Help students use this medium to create images using words and sounds.
- Provide a noise-free environment for recording. Emphasize the importance of eliminating all unwanted sounds. (Even the sounds made by turning the pages of the script may be picked up.) Have the students listen carefully to their preliminary tape to avoid noise problems in the final recording.
- Discuss how these procedures relate to radio production.

VIDEO TAPE

- Discuss the elements of the video medium that must be considered.
- Present concepts of storyboarding (content, images, script, and sound effects) as a plan. Storyboard should be evaluated by students and teacher before actual shooting script is prepared.
- Explain how script blanks (sample provided) may be used. (Follow explanation included in the Procedure sheet.)
- Provide a noise-free environment for recording. In the context of video, "noise" means not only unwanted sound, but also unwanted visual images (foreground, background, shadows, etc.). Emphasize the importance of eliminating all unwanted noise. Have students watch and listen carefully to their preliminary tape to avoid noise problems in the final video.
- Discuss how these procedures are related to television production.

Answers to Review Questions

1. Discuss the roles of the seven technological resources in the completion of this activity.
   Answer: People - content selection, writers and artists, production
   Information - about the audience, production techniques, content
   Materials - paper, audio tape, video tape
   Tools and machines - computers (input, process store, output data), audio recorders, video cameras/recorders
   Energy - electrical (power, transmit, receive), magnetic (data storage), light
   Capital - to finance activity
   Time - person-to-person communication, machine communication, person-machine communication

2. How did the three parts of this activity attempt to inform, persuade, educate, and entertain?
   Answer: The content of the articles/stories in all three media may have had the goals of informing, educating, and/or entertaining. In addition, the advertisement(s) were probably attempting to persuade the audience.

3. What information is provided by a rough layout?
   Answer: Information provided by a rough layout includes: paper size, image area (margins), location of images (text and graphics), image sizes (text and graphics), approximate type sizes and styles, and wording of display heads.

4. Discuss the advantages of desktop publishing and manual paste-up procedures.
   Answer: Advantages of desktop publishing: Ability to compose entire pages, including heads, text, and graphics. Utilization of computer graphic capabilities. WYSIWYG (What You See Is What You Get) capability allows for arrangement of text and graphics on the computer screen exactly the way you want them to print on paper.

   Advantages of manual paste-up: Easy utilization of different types of output. No special computer equipment necessary. May be done at any site.

5. In producing your newsletter, what methods were used to create graphics and to convert photographs to halftones?

    Answer: Methods use to create graphics include computer "draw" and/or "paint" programs, computer graphic programs, hand drawing, digitization (scanning), and commercially prepared art: 1. clip art (preprinted) and 2. click art (on disk).

    Methods used to convert photographs to halftones include conventional (photographic) halftone conversion process and digitizing (scanning).

6. List three methods that could have been used to input data for your newsletter.

    Answer: Data may have been input using one or more of the following methods: keyboard, floppy disk, hard disks or tapes, optical character reader, digitizing tablet, speech recognition, and scanner.

7. What roles did electronic technology play in completion of the following activities: newsletter, audio tape, video tape?

    Answer: Electronic technology that played important roles in completion of the three parts of this activity included data communication, data input, data processing, data storage and retrieval, computer output, and power.

8. What is "noise" in relation to audio and video? Why is it an important factor to consider?

    Answer: In relation to audio and video, noise is unwanted sounds or images. It is important to eliminate noise in the final production because it interferes with effective communication of the message.

9. What are the similarities and differences between a rough layout (newsletter) and a script (audio and video)?

    Answer: Similarities: Both a rough layout and a script provide an overview of the final product, including a plan, parameters, wording, and graphics.

    Differences: A rough layout may not specify the exact wording (text) of all of the stories.

10. What advantages and/or disadvantages did you discover in communicating using each of the following three media: (1) print, (2) audio, and (3) video?

    Answer: The answer to this question will be based upon the specific nature of the experiences.

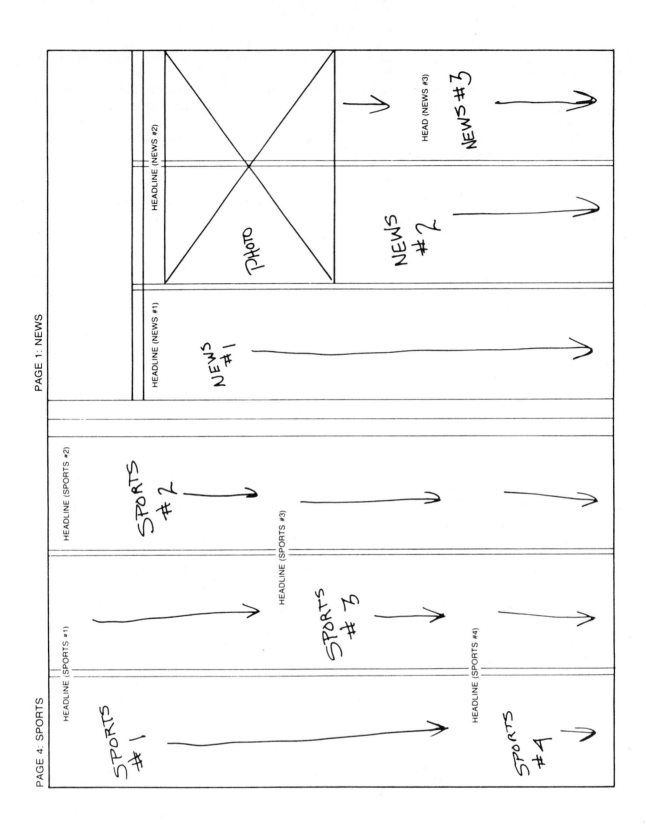

PAGE 1: NEWS

PAGE 4: SPORTS

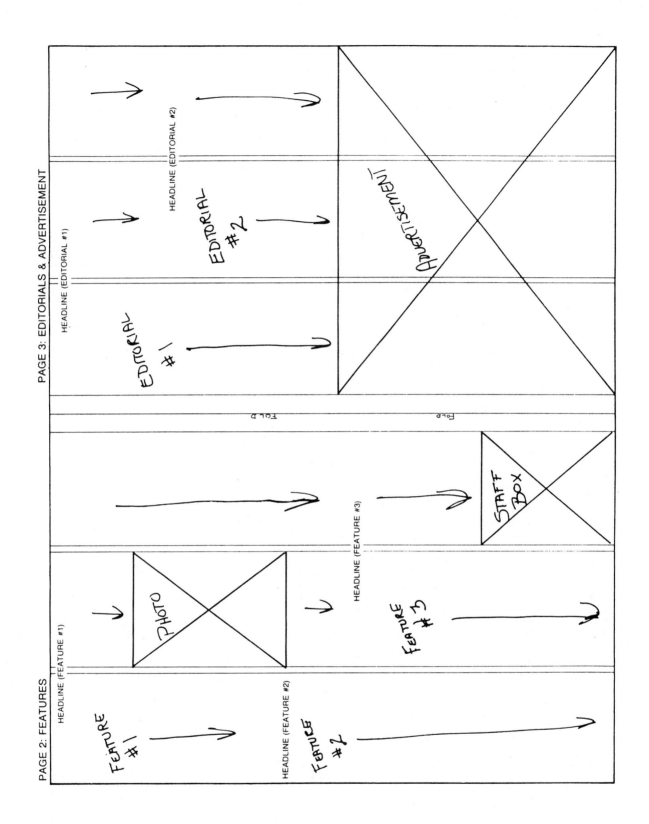

397

TITLE: _____

PAGE _____ OF _____

VIDEO	AUDIO

Sample script blank

CUTTLEBONE CASTING

Suggested Activity Length – 2 to 3 weeks

Activity Objective – In this activity, the students will learn about casting as it relates to jewelry making.

Instructor Guidelines – Using cuttlefish bone, the student will create a pattern for making a piece of jewelry. Make sure that emphasis is placed on safety when handling the molten metal.

If you are unfamiliar with this procedure, you will want to experiment before doing this activity with the class. Care should be taken when creating the mold, for when handling a small item it can easily get broken.

Be sure to obtain the correct metal so that several castings may be made. Use of an imitation mixture may produce an imperfect casting.

Answers to Technology Connections

1 What is an alloy? What metals are in modern pewter? Older pewter is an alloy of what metals?
   Answer: An alloy is a metal which is composed of two or more metals. Modern pewter is an alloy having tin as its chief component. Older pewter is an alloy that contained lead.

2. Why have lead-based products been changed to lead-free products?
   Answer: Lead has been found to have harmful medical effects on the body when ingested.

3. Identify the forming, combining, and conditioning processes used in making your jewelry.
   Answer: Casting, shearing, drilling, grinding, shaping, heat treating, polishing.

CHECKERBOARD

Contributed by Thomas Barrowman

Suggested Activity Length – 2 to 3 weeks

Activity Objective – In this activity, students will process natural and synthetic materials into a three-dimensional game board.

Instructor Guidelines – Frame a few sample boards and test the casting procedure before attempting to cast student projects. Thickness will affect the resin shrinking, so test the thickness to be cast.

It is important to use the manufacturer's recommended amount of catalyst. This varies according to the thickness to be cast.

The Industrial Arts Supply Company, 5724 West 36th Street, Minneapolis, MN 55416 can supply you with clear polyester casting resin.

Experiment with casting a thin final top layer of casting resin against a waxed plate glass, or use mylar between the glass and resin.

By using 9 strips of material instead of 8 strips, the strips will not have to be reversed in order to make the checkerboard. This process will produce a more accurate board. Do not be discouraged if each square does not line up perfectly with the next square. Look carefully at professionally-made boards and you will find that they are not in absolute alignment.

Answers to Technology Connections

1. Mylar will release from polyester resin. What other release agents might you use?
   Answer: Carnauba waxes, polymer protective coatings (used in place of wax), silicone spray, and teflon spray. Ask the company where you buy clear casting resin what type of materials can be used as a release. New products are always being developed.

2. Identify all the tools and machines used in this lab.
   Answer: Circular saw, power sander, jointer, try square, bench rule, miter clamps, bar clamps.

3. Identify each material used as to whether it is renewable or nonrenewable.
   Answer: Wood and paper are renewable. Glues, polyester resin, catalyst, abrasive papers, and waxes are nonrenewable.

4. List the procedures of material conversion that were used in this activity. These conversion processes can be listed under forming, separating, combining, and conditioning.
   Answer: Forming processes - casting
   Separating processes - sawing, sanding
   Combining processes - gluing wood strips and assembling the picture frame around the checkerboard, making a composite with wood and polyester resin, applying release agents and final waxes.

AN ENTREPRENEURIAL COMPANY

Contributed by A.R. Putnam

Suggested Activity Length – 2 to 3 weeks

Activity Objective – The purpose of this activity is to provide a problem-solving activity to present product design concepts. Students work in groups of 4 or 5 persons to create an original and practical design. Brainstorming and originality should be stressed.

Instructor Guidelines – Design parameters can be adjusted to suit material and equipment availability.

You may wish to set the stage somewhat for the presentation of the products. The presentations work best when they are formal and well prepared. Students should be reminded to judge designs based upon objective criteria. Positive features should be pointed out for all designs. Remind the students that the company can only choose one design.

Answers to Technology Connections

1. Quality circles allow workers to become directly involved in decisions about the production process. What effect has this had on the workers and the companies?
   Answer: The workers do a better job and the company becomes more productive.

2. Many of the products you use every day are manufactured by machines and automated equipment. How would your life change if everything you use had to be made by hand?
   Answer: Products would be much more expensive. We would not have the variety of products we have now.

3. New production systems make increasing use of technologies such as robotics, CAD/CAM, and CIM, just to name a few. How will this affect you as a future worker and consumer?
   Answer: More education is needed before entering the world of work; products will be more varied and less expensive.

T-SQUARE ASSEMBLY LINE

Contributed by Alan Horowitz

Suggested Activity Length – 1 to 3 days

Activity Objective – In this activity, you will set up an assembly line with your class in order to demonstrate some of the mass production techniques used by industry. In this case a basic drawing tool, known as a T-square, will be manufactured.

Instructor Guidelines – You may wish to use this activity early in the semester to break up the long series of introductory lectures (i.e., course requirements, safety rules, measurement, drawing and planning, etc.) commonly presented at the beginning of a course.

This T-square activity can be included as part of a series of drawing and planning activities focusing on the evolution of graphic communication. Students should first draw an object freehand. Next, the students should use the T-square and other drawing tools to produce the same drawing. Finally, students should use a computer-aided drawing (CAD) system to again draw the same object.

The jigs and fixtures, which speed the assembly line by making it more efficient, can be constructed by the entire class as a problem-solving activity or you may wish to construct them in advance. The entire activity will usually take two or three days after the jigs and fixtures are made.

All the lumber should be brought to the proper widths and thicknesses in advance using a planer, table saw, and jointer. The edges of all the pieces should be straight and smooth.

The speed and efficiency of your assembly line will be determined by many things, including the layout of your room and the number of machines and tools (miter boxes, drills, assembly jigs, etc.) used with your class.

Portable pieces of machinery and equipment will usually increase the flexibility of a technology facility, especially when used with mass production/assembly line activities.

The number of students in your class can be a problem. If there are not enough jobs to go around, you can have the overflow students do time studies, take photographs or finish sanding. If your class is very small, students can switch jobs to keep the assembly line process moving.

The power screwdrivers and/or variable speed electric drills with driver bits are very important to the efficiency of the assembly line. Brass Phillips head screws work well but are expensive and difficult to find. If flat head screws are used in place of the round head ones, the shank holes drilled in the T-square blades will need to be countersunk.

Do not use glue when assembling the T-squares. Glue will prevent any adjustments from being made after the drawing tool is completed. If any of the T-square heads or blades are nicked, they should be rejected and repaired. Ask your students if they would buy such a damaged product in a store. Insist on high-quality results!

If desired, a 6" plexiglass 45/90-degree triangle can be mass produced for use with the T-square. Use 1/8" clear plexiglass for the triangles and cut the angle on a miterbox. Polish (<u>do not</u> round) the edges with very fine abrasive paper. Drill a 1" finger hole in the center of the triangle using a drill press.

Answers to Technology Connections

1. What are some advantages of CAD?
   Answer: CAD is faster; easier to correct, edit or change scale; drawing storage is more convenient (i.e., on a disk or tape); drawings can be sent via a phone line with a modem, etc.

2. Why is accuracy so important in designing and making jigs and fixtures?
   Answer: If the jigs and fixtures are not accurate to begin with, the products produced using the jigs and fixtures are also likely to be inaccurate.

3. Why are quality control checkpoints important in an assembly line? Interchangeable parts?
   Answer: Without quality control checkpoints in an assembly line, the finished product may not meet the minimum acceptable standards for accuracy, safety, or marketing, The repair of an inferior product can cost more than the continual maintenance of quality control checkpoints. Interchangeable parts are a necessity for the efficient mass production of a product on an assembly line.

4. There are two basic subsystems of the manufacturing system: (1) material processing and (2) business and management. Which subsystem does developing an assembly line flowchart fall into? Marketing? Final assembly?
Answer: An assembly line flowchart and the final assembly of a product usually fall into the materials processing system. Marketing is usually a function of the business and management system.

5. Why is it important to use standardized measuring systems in manufacturing?
Answer: Standardized measuring systems, like interchangeable parts, are necessary for the efficient mass production of a product on an assembly line as well as the ability to manufacture and market products on a national and international scope.

ROBOTIC RETRIEVER

Contributed by Grant Luton

Suggested Activity Length – Study pp. 16–17 of the robot manual, and experiment with the commands until you are familiar with each one; 45 minutes. Graph the room. Write, test, and debug the program; 45 minutes.

Activity Objective – Students will successfully program a robot to perform a specific task.

Instructor Guidelines – It is not necessary that steps #1–3 be performed by each class. This may be done by the instructor ahead of time. On the other hand, a group of students may construct a room of their own design as a challenge for another group.

Student handouts may be prepared ahead of time. These may provide places for student names, a graphing area, program flowchart area, and any other information you may wish the students to record.

This activity works well with a group of students. If several Mode 603s are purchased, then enough stations could be set up to accommodate the entire class. Graymark also manufactures an entire line of moderately-priced robots which could provide alternative activities for several groups. Write or call Graymark International Inc., Box 5020, Santa Ana, CA 92704 (1-800-854-7393).

Answers to Technology Connections

1. Robots became practical only after the invention of the computer. Can you name what tasks are performed by the computer inside the 603 robot?
Answer: Memory (to store the program), processor (to turn ON and OFF the motors, light and horn).

2. More advanced robots are able to sense certain things in their environment. What kinds of senses would make your robot more useful?
Answer: Sensitivity to magnetism, radioactivity, heat, light, touch, etc.

DOME CONSTRUCTION

Contributed by Alan Horowitz

Suggested Activity Length - 2 to 4 weeks

Activity Objective - In this activity, students will assemble a dome-shaped structure in order to demonstrate the principle of an alternative design and construction technique.

Instructor Guidelines - CAUTION: The pentagon and hexagon shapes that are cut out on the notcher for this activity greatly resemble "Chinese stars" that students see in the "kung-fu" movies. These can be very dangerous if they are thrown through the air at high velocity. If you think this will be a problem with your class, make the pentagon and hexagon shapes out of another material or, with a slight design change, the shapes can be made into circles. Some of the circles will have 5 holes and some will have 6 holes punched in them.

CAUTION: Do not allow students to drill holes in the sheet metal with the drill press or power drill unless properly clamped. Sheet metal has a tendency to grab and spin loose.

Domes can be constructed of many different kinds of material including soda straws and pipecleaners, toothpicks, folded paper, strips of wood, plastic, glass, and metal. The dome design suggested in this activity is only one of many ways you may wish to approach this topic.

Because there are a large number of parts in the dome structure that need to be cut on the notcher, it would be helpful to have more than one notcher available in the lab. This is also true for the hole puncher.

You should duplicate the fuller size shapes A, B, and C shown in the text and allow students to use them as templates for making these dome parts.

This activity can be done as a mass production project. Jigs and fixtures and stops can be made from sheet metal and attached to the notcher with masking tape and/or clamps.

Do not completely tighten any of the screws until the entire structure is assembled. A nut driver will make tightening the 100+ screws easier.

The dome can be covered with clear plastic stretched over the superstructure and fastened with duct tape. Make sure there are no sharp points on the surface that can puncture the plastic.

Answers to Technology Connections

1. A simple dome is not hard to build. It needs no columns or supports to hold it up. Why not?
   Answer: The weight of a domed structure is transmitted uniformly from its top outward and downward to its base. As a result no internal weight-bearing supports are needed.

2. How is the heat from the sun concentrated by a greenhouse?
   Answer: The short-wave radiation from the sun penetrates the glass or plastic covering of the greenhouse, warming the surface inside. The longer-wave radiation being radiated back from the surfaces cannot go through the glass or plastic and is trapped inside.

3. Why is it possible to use smaller foundations with dome structures?
   Answer: Domed structures are comparatively light in weight, therefore the footings and foundations for this type of structure can be significantly smaller than those used with more traditional construction techniques.

4. Which two geometric shapes are usually part of the design of a dome? Are there any more?
   Answer: Hexagons, pentagons, and triangles.

5. What materials can be used to build a dome? Is an igloo a dome?
   Answer: All sorts of materials can be used to make a dome including paper and cardboard, metal, wood, plastic, and even ice or snow. An igloo is considered a dome.

BRIDGE CONSTRUCTION

Contributed by Alan Horowitz

Suggested Activity Length – 2 to 3 weeks

Activity Objective – In this activity, students will explore the engineering and construction problems involved in designing and producing a bridge/supporting structure.

Instructor Guidelines - This activity can be done using many variations. Toothpicks, straws, paper, smaller dimension strips, balsa wood, etc. can be used to assemble the bridge with or without gussets. The support weight requirement can be varied to meet the construction materials being used.

By using a small strip (8"-12") of the wood being used to construct the bridge as a guide, the full-scale drawing of the bridge goes much easier. Students can lay the piece of wood on the graph paper and use it as a template. Use a very fine-toothed saw to cut the strips of pine to length. A hacksaw works well.

You can test the bridges by simply stacking weights on the structures. You can also use an arbor or hydraulic "press" (a car jack is fine) and a bathroom scale to apply pressure and measure the stress weight. Also, a large bucket can be placed on the bridge. This bucket is slowly filled with water or sand. If you use water, do the testing outside or in a sink and be prepared to get wet.

To prevent complete destruction during the testing phase, "stops" or safety blocks can be cut and placed in strategic locations between the stress weight and the bridge. If the bridge gives way, the blocks prevent total destruction by supporting the weight. They will not allow the structure to be crushed flat.

To make the reinforcement gussets for the structures, have the students use a ruler to make a grid pattern (3/4"-1" squares) on a file folder or card stock. Cut the grid squares out and then cut these diagonally in half. These triangles are then used to reinforce the corners of the bridge.

Some students may want to have a contest to see which of their structures will hold the most weight. A well-designed bridge can hold a lot of weight, so be careful that nobody gets hurt if the bridge structure suddenly collapses.

If a bridge collapses during testing, ask the "junior engineer" to locate a point or points of failure. See if the class can explain why the structure gave way to that specific point and, as a group, determine what could have been done to prevent the failure.

Answers to Technology Connections

1. Construction refers to building a structure on a site. What factors must be considered when choosing a site for a bridge?
   Answer: Stability of ground at the site location, width of the span, accessibility to the structure, etc.

2. What type of bridge has its deck supported from above by steel cables?
   Answer: A suspension bridge.

3. What construction material is made from a mixture of cement, sand, stone and water?
   Answer: Concrete.

4. Engineers must calculate all of the stresses (forces) on bridge structures they design. Can you pick out some of the major stress points on your bridge?
   Answer: Corners, joints, and surfaces where there is tension and/or compression of the structural material.

5. What can happen to a bridge if it is not engineered properly? Has this ever happened?
   Answer: It can deteriorate, be unstable, and, in some cases, collapse.

PREFAB PLAYHOUSE

Contributed by Thomas Barrowman

Suggested Activity Length – 2 to 3 weeks

Activity Objective – In this activity, students will learn about working together as a team to produce a product. Contracts, record keeping, and research will be part of the activity, along with building construction.

Instructor Guidelines – By substituting different sizes of studs, rafter, and sheathing, you will be able to weatherize the playhouse for different areas of the country.

Carriage bolts should be used for the assembly of the wall sections. Different corner assemblies can be constructed with the use of 4" x 4" boards and carriage bolts.

You could easily expand this activity and produce storage sheds using this prefabricated method of construction. Construct a rafter assembly jig using four 2" x 4" boards and nailing spacer blocks to the surface of the 2" x 4".

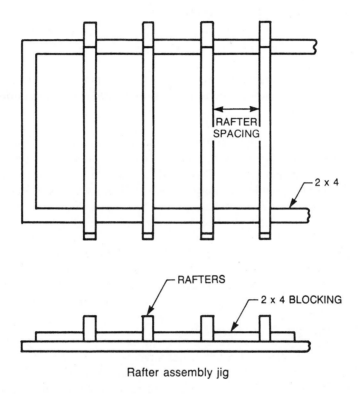

Rafter assembly jig

Lay each precut rafter in the assembly jig with the bird's mouth facing up. Nail a ridge board in place and a top plate in the bird's mouth.

Carefully turn the unit over and place it back in the jig. The rafters will now be facing in the correct direction so you can nail your covering material in place.

Answers to Technology Connections

1. You need to supply the purchaser with a snow-load specification for the roof you designed. You will find this information and charts in the reference book you used. Specifications for snow load or live load will vary according to the climate and building codes. Use the specifications for your area.

2. Describe some ways of making your roof stronger.
   Answer: You can make the spacing between the rafters shorter, double the number of rafters, use a larger rafter, install collar beams, or use a truss.

3. What are the advantages and disadvantages of on-site construction?
   Answer: Advantages of On-Site Construction: Individual customization, being able to take advantage of lower current prices, more options for the customer, being able to choose a subcontractor.

   Advantages of Prefabricated: Not having to deal with the weather, not having to deal with late deliveries, better control of safety in construction, faster speed of completion, more uniformity of materials.

4. Why are studs usually placed 16" or 24" on center for framing?
   Answer: The 4' x 8' wall board or paneling must end on a stud where there is support.

INSTALLING ELECTRICAL SYSTEMS

Contributed by A.R. Putnam

Suggested Activity Length – 1 to 2 weeks

Activity Objective – In this activity, students install a simple house wiring circuit simulating the way it would actually be done in the residential construction industry. Students should work in groups of from 2 to 4 students per circuit.

Instructor Guidelines – Safety tips: Copper wires are sharp, so everyone must wear safety glasses with side shields at all times. Never allow a circuit to be connected to a live source until you have checked it. You may wish to turn the power off in the lab, just to insure safety. No part of the circuit should be touched when it is live. Students should be warned that they should never work on a live or unknown circuit in the lab or at home, as normal house wiring can deliver a lethal shock!

House wiring circuits can be easily varied in complexity. You may wish to assign additional circuits to more advanced students.

Terminals on the electrical appliances can be damaged by rough handling. Reasonable care should be exercised when placing the heavy wire around the terminal and removing it.

Copper wire hardens when it is worked. When students first cut lengths of wire, encourage them to be generous, so that enough wire will be left to cut off the hardened tips when the length is reused.

Answers to Technology Connections

1. An electrical circuit in a structure is made up of several devices which have different purposes. What is the purpose of a duplex outlet receptacle? A switch? A light fixture outlet?
   Answer: To allow electricity to be used by an appliance; to interrupt a circuit; to provide a means of supplying electricity to a light bulb.

2. Many of the things in our homes work by electricity. How would your life be changed without electricity in your home?
   Answer: No electrical lights, television, stereo, electric heat, air conditioning, refrigerators, etc.

3. What happens when someone tried to use more electricity in a structure than the electrical system was designed to provide?
   Answer: The fuse blows or the circuit breaker trips.

LASER CONSTRUCTION

Contributed by Thomas Barrowman

Suggested Activity Length – 1 to 2 weeks

Activity Objective – Students will learn how to safely use the laser in three different building applications.

Instructor Guidelines – Drain Pipe Alignment: Several jigs similar to the one below have to support and adjust the drain pipe when the activity is to be performed in the room. Students should be warned to use laser safety glasses when working with a laser.

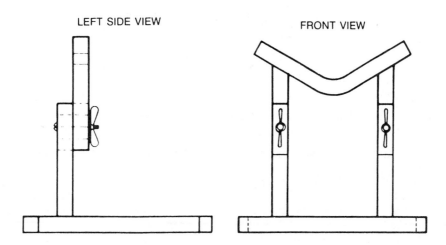

LEFT SIDE VIEW   FRONT VIEW

Concrete Blocks: The laser can be used to lay real concrete blocks. Keep in mind that the blocks should be laid out relative to the scratched reference line and not necessarily exactly on the reference line.

You may want to make several jigs so that construction can begin at different places on the wall. By making a jig with 4 reference points on the acrylic, you can lay more courses of blocks without having to readjust the laser.

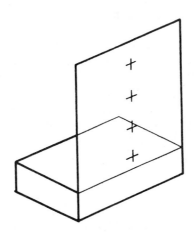

Answers to Technology Connections

1. Using a 4-foot level, tape measures, and chalk lines, try accomplishing the same tasks you have just accomplished using the laser. Once the student becomes familiar with the laser setup, he or she will find it to be much faster than using chalk lines.

2. Could you have the laser turn a corner by using a mirror?
   Answer: Yes, the mirror would reflect the laser light, causing it to "bend."

3. List the operations that have to be accomplished before a foundation wall is built and aligned with a laser.
   Answer: Plans must be submitted for a building permit, foundation hole dug, batter boards installed, building lines run, footing trench dug, footing forms constructed, and footings poured.

4. Why are alignment and leveling important in the building of a structure?
   Answer: If the alignment or leveling of one part of a structure is off, the whole structure would be "off." Therefore, the structure possibly would not be safe.

ONE, IF BY LAND

Contributed by Alan Horowitz

Suggested Activity Length – 2 to 3 weeks

Activity Objective – In this activity, you will set up a small company and assembly line with your class in order to expose students to business management and mass production techniques.

Instructor Guidelines – NOTE: Make enough extra lantern parts to compensate for the "rejects" quality control will kick out.

The jigs and fixtures, which speed the assembly line by making it more efficient, can be constructed by the entire class as a problem-solving activity, or you may wish to construct them in advance. In many cases, the jigs and fixtures used in this activity can be pieces of scrap metal or bench rules clamped to the machines and used as stops or guides.

The speed and efficiency of your assembly line will be determined by many things, including the layout of your room and the number of machines and tools (squaring shears, notchers, box and pan brakes, bar-folders, etc.) used with your class.

Portable pieces of machinery and equipment will usually increase the flexibility of a technology facility, especially when used with mass production/assembly line activities. The spot welder is important to the efficiency of the assembly line. Soldering could be used instead of the welder, but would slow the production line.

If any of the lanterns are poorly constructed, they should be rejected and/or repaired. Ask your students if they would buy a damaged product in a store. Insist on high-quality results!

Answers to Technology Connections

1. Why is it important to make a prototype, or "original" working model, of the project being mass-produced before going into full production on the assembly line?
   Answer: Many of the potential production line and design problems can be located and solved in the prototype stage. A prototype can go through many changes before it reaches the final production line stage.

2. Why is accuracy important when manufacturing most products?
   Answer: Accuracy is important to ensure that the product being manufactured meets acceptable standards for safety or marketability.

3. What is the purpose of quality control checkpoints?
   Answer: Frequently-located quality control checkpoints can usually spot any defective products on the assembly line more quickly.

4. How do mass production techniques differ from the craftsman techniques?
   Answer: Mass production techniques usually use minimally-skilled workers doing small steps in the manufacturing process or on an assembly line. The craftsman was a highly skilled individual who usually manufactured an entire product alone.

5. Why is advertising so important to the manufacturing industry? Make a list of the different ways manufacturers can advertise a product.
   Answer: The success of many manufacturing industries depends on the competitive marketing of its products. Advertising is frequently used by industry to sell these products. In a very competitive marketplace, an effective advertising campaign can substantially increase the profits of a company. Newspapers, magazines, television, radio, billboards, shopping bags, Yellow pages, Goodyear blimp, sky-writing, bumper stickers, T-shirts, etc. are advertising vehicles.

6. Match the advertising slogans:

1. Soup is good food!	Allstate Insurance	6
2. 99 and 44/100% pure	New York City	9
3. The wings of man	Campbell's Soup	1
4. When it rains, it pours!	General Electric	5
5. We bring good things to life!	U.S. Army	8
6. You're in good hands...	Greyhound Bus	10
7. Over 30 billion sold	Eastern Airlines	3
8. Be all that you can be!	Ivory Soap	2
9. The Big Apple	MacDonald's	7
10. Leave the driving to us!	Morton Salt	4

LANTERNS, INC.

JOB APPLICATION

Name _____   Date_____

Address_____   Phone Number_____

_____   Social Security # _____

For what position(s) do you wish to apply?_____

Salary Desired?_____

Approximately how many days were you absent from work last year ?_____

Highest education degree/diploma held_____

Do you have any specialized training or experience related to the position for which you are applying?_____

If yes, where and when did you get this training or experience?

_____

_____

Give three references:

Name	Business Address	Telephone	Profession

Why do you think you would make a good candidate for this position?

_____

_____

_____

_____

For the interview use only. 1 (excellent) -> 5 (poor)

Appearance_____   Education_____   Personality_____
Experience_____   Attendance _____   Salary Desired_____

TOTAL_____   Recommendation:_____

PRODUCTION COMPANY

Contributed by Doug Polette

Suggested Activity Length – 5 to 15 days

Activity Objective – In this activity, the student will design and engineer components to be manufactured in the classroom company and then construct a structure made from these components. Furthermore, the student will develop an understanding of geometric shapes as well as the inputs, processes, outputs, and feedback sections of the universal systems model.

Instructor Guidelines – Overview Activities: Draw on the board or make an overhead of the universal systems model and explain, discuss, and provide examples of each part.

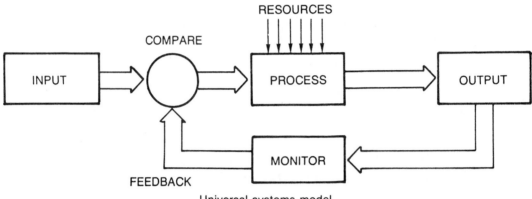

Universal systems model

Make sure the student has an understanding of the overall concept of production, including both construction activities (those production activities which require work to be done on site) and manufacturing activities (those production activities which allow the work to be completed in a plant). Naturally the planning, designing, engineering, etc. activities precede the actual work activities.

Setting up a Company: Set up a production company as outlined in the activity. Explain the relationship between the different parts of the company. Provide local examples and, if available, local speakers.

If a computer is available for classroom use and has a CAD program, allow the students to design their components on the computer. Then demonstrate the ease with which they may change the original drawing. Discuss the various ways to obtain capital for the operation of the company, such as a single proprietorship, a partnership, or a corporation. Discuss the advantages and disadvantages of each.

Producing the Parts: The students may encounter problems when trying to figure out how to connect the ends of the components to each other, particularly when these connections are not in a single plane but at various angles. A few ideas are: the use of small sheet metal angle brackets, hot glue, silly putty, small bolts, paper clips, etc. If possible, let the students solve this problem on their own; you might be surprised at their innovativeness.

Make sure prototypes are built to determine overall looks, appropriate materials, and strength before actually starting on the design of the various components.

Carefully monitor the design and construction of the jigs and fixtures to determine the following:

1. Safety in the use of the jig.

2. To maintain the greatest accuracy, dimension all parts from one end and one side.

3. Provide clamping devices to hold stock into position.

4. Make sure jigs and fixtures will produce parts within acceptable tolerances.

5. Build test devices to test the size and hole location of the components to assure the ease of assembly of the structure.

6. Make holes, slots, etc. oversize, if possible, to provide an extra margin for error.

7. For the actual run, move equipment and tools into position to reduce the amount of transportation of parts from one machine to another. Let the students determine the best location of the equipment for their production run. Adjustments may have to be made due to size of the equipment, electrical power, air, water, etc. In this case, transportation devices can be designed to make the production run efficient.

8. Make sure, after all equipment is in position with all jigs and fixtures installed, that a trial run is made before the actual production run takes place. Adjustments are easy to make during the trial run but very difficult during the actual production of parts.

9. Plastic bags, sacks, small boxes, cans, etc. are very handy for storing parts. Make sure to label all containers with the proper identification so parts will not get mixed up and a proper count can be taken.

10. If you are operating more than one class with this activity, it may be desirable to have one class provide assembly directions to another class as a check to determine their clarity and completeness. This is also an excellent opportunity to team with the language teacher to help the students write clearly.

Answers to Review Questions

1. List the seven resources that are required by any production company in order for it to be successful.
   Answer: People, information, capital, tools and equipment, time, materials, and energy.

2. List several inputs of the engineering and design division of your company. Do the same for your manufacturing division and your construction division.
   Answer: Engineering and Design Division – Ideas, time, drawing equipment.
   Manufacturing Division – Drawings and designs from that division, and the other inputs listed in Question #1.
   Construction Division – The output from the manufacturing division plus resources listed in Question #1.

3. What is the processing phase of the manufacturing and construction divisions?
   Answer: The actual production activities of the manufacturing and construction division.

4. List the outputs of the manufacturing and construction divisions of your company.
   Answer: Outputs of the manufacturing division are the parts produced. The output from the construction division is the completed structure. Outputs also include unexpected and undesirable results, such as loss of jobs to automated manufacturing.

5. Determine those activities that provide feedback for your company.
   Answer: The feedback for the manufacturing division would be how well the parts meet the specifications and drawings. For the construction division, the feedback will be how well the structure meets the original structure design and how well the directions were written for building the structure.

6. Discuss what could happen to a company if any one of the employees did not do the best possible job.
   Answer: The entire company's productivity would decline.

7. Make suggestions on how the manufacturing division of your company could have produced the components more efficiently.
   Answer: Will vary

8. Make suggestions on how the workers in the construction division may have been able to do their jobs more efficiently.
   Answer: Will vary

9. Study your surroundings and list at least ten different items that are built in such a way that a triangle is used in the design to add strength and stability to the device of a structure.
   Answer: Will vary

10. List the major differences between a constructed item and one that has been manufactured.
    Answer: Generally a manufactured item is one that is built in a plant; a constructed item is one that is built on site.

HOT DOG!

Contributed by Alan Horowitz

Suggested Activity Length – 3 to 5 weeks

Activity Objective – In this activity, students will construct a device that can cook a hot dog and thus demonstrate the use of the sun as an energy source.

Instructor Guidelines – Refer to the attached drawings for the steps in laying out a parabola with a 6″ focal length. Draw the layout on a large sheet of 1/4″ square graph paper. Draw the X and Y axes (with "O" being the intersection), and mark point P1 and the focal point (F). Use a large compass or trammel points to mark and plot the parabolic shape. Measurements A = A1, B = B1, C = C1, etc. The more points you make, the more accurate the parabola. Connect the plot points with a flexible curve. The accuracy and size of the parabola will greatly affect the operation and efficiency of the cooker.

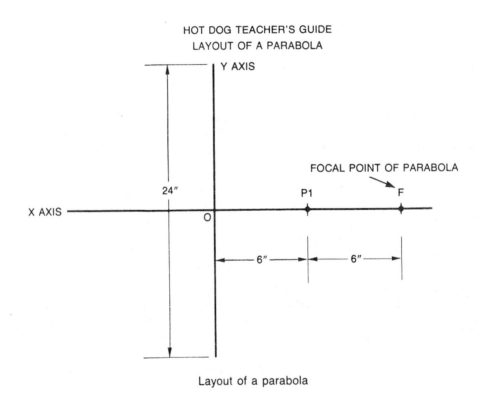

Layout of a parabola

On a bright sunny day, a hot dog will usually cook in 15–20 minutes. If the hot dog is wrapped in black aluminum foil, it will cook much faster.

If the mirrored mylar is cemented to the tin plate before it is curved into the parabolic shape, it has a tendency to wrinkle and blister when the tin plate is bent into final shape. If possible, run the tin plate through a slip-roll former to slightly curve the metal before adhering the mylar to its surface.

Remind students that if the mylar reflective surface becomes dirty, it should be gently cleaned with nonabrasive soap and water and a sponge or soft cloth. If this surface becomes badly scratched, it should be replaced with a new sheet of mirrored mylar.

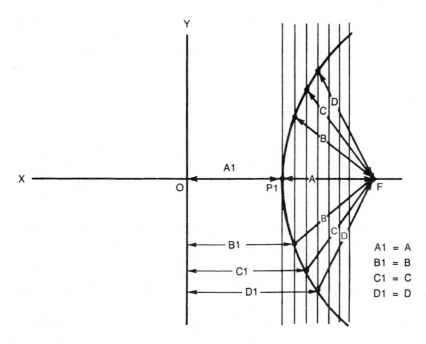

Layout of a parabola

Answers to Technology Connections

1. What are some advantages of using solar energy? What are some disadvantages?
   Answer: Advantages – Solar energy is free, nonpolluting, very versatile, renewable, etc. Disadvantages – Only good in the daytime, clouds are a problem, can be expensive to harness, etc.

2. How is the heat from the sun concentrated by the solar cooker? Why must the parabolic reflector be as smooth and reflective as possible?
   Answer: The sunlight is concentrated by the parabolic mirrored surface reflecting most of the light to a single area called the focal point.

   The higher the reflectivity of the surface of the parabolic mirror, the more efficient the solar cooker will become.

3. White objects (like marshmallows) do not cook very well on the cooker. Why not?
   Answer: White surfaces reflect the sunlight. Dark objects absorb the light and will heat up more quickly.

4. What effect would increasing the size of the reflector have on the time it takes to cook a hot dog? Why?
   Answer: Everything else being equal, the larger the reflector, the faster the hot dog will cook. The larger parabolic reflector will capture and concentrate more of the sun's energy.

A PENNY FOR YOUR THOUGHTS

Contributed by Alan Horowitz

IMPORTANT: Inform students that if they have not been instructed in the proper and safe use of any tool or machine that they need to use, check with you <u>before</u> going any further.

Suggested Activity Length – 2 to 3 weeks

Activity Objective – The purpose of this activity is to provide a problem-solving activity that graphically demonstrates the difference between potential and kinetic energies. The penny toss activity can be done individually or in teams of two or three students.

Instructor Guidelines – Some of these penny toss devices can be quite powerful. Safety glasses must be worn at all times. Students must be careful not to shoot their devices at other students. A hacksaw can be used to cut the thin wooden strips to length.

Instead of using hot glue, white glue used with gussets made from card stock work well. Pour small amounts of the glue in an empty 35mm film cartridge and use a coffee stirrer to do the spreading.

You may wish to judge students on their originality, neatness, and construction techniques as well as accuracy. Also, students who use the least amount of material can be given bonus points. A competition for the greatest throwing distance can also be held. As 67'3" is the record so far, you may have to hold this event outdoors or in the gymnasium.

Answers to Technology Connections

1. Solving technological problems requires skill in using all seven resources. These resources are people, information, materials, tools and machines, energy, capital, and time.

2. Where did your penny-toss device get its energy?
   Answer: From the potential energy you placed in the rubberband when it was stretched.

3. What types of devices were used in the Middle Ages to throw or shoot stones or arrows?
   Answer: Catapults, long bows, and crossbows with arrows.

4. What do you think would happen to the throw distance or range of the penny-toss device if you used something heavier than a penny?
   Answer: When the mass of projectiles is increased, the range of the device usually decreases.

5. What can you do to increase the range of the penny-toss device?
   Answer: Add more rubber bands to increase the total potential energy, cut down on friction in the moving parts, increase the control, angle, or trajectory of the penny, etc.

6. Why does the penny continue to fly after it has left the penny-toss device? (Inertia)
   Answer: Newton's first law of motion: If an object is at rest, it tends to stay at rest; if it is moving, it tends to keep moving at the same speed and in the same direction.

MUSCLE POWER

Contributed by Fred Posthuma

Suggested Activity Length – 2 to 4 class periods

Activity Objective – After completing this activity, the students should understand the terminology involved and the process of changing potential energy of their muscles to kinetic energy to electrical energy produced by the bicycle and generator. The student should also understand that technology extends human potential energy through the use of machines.

Instructor Guidelines
1. Introduce the activity as a class competition, possibly with prizes.
2. It takes some time to construct the bicycle/generator, so plan ahead. It might be possible for an advanced student to complete the actual construction.
3. Emphasize the concepts of energy, work, power, and energy transmission.
4. A recording sheet for each student should be used to keep track of results.
5. Most students will get tired quickly after lights are turned on due to the loading of the generator. This means you must take results and measurements fast before students tire. Take their highest readings.
6. You may also want to introduce the concepts of voltage and current.

Bicycle/generator set-up: See text pages 398 and 399 for diagrams of the bicycle/generator set-up and the wiring for the load/test components.

Helpful hints on construction:
1. You may use a single- or multi-speed bicycle. Obviously more concepts such as gear ratios and mechanical advantage could be covered with a multi-speed, but a single-speed will be easier to find, cheaper and sturdier.
2. A generator is better to use than an alternator. Most salvage yards will have a 12-volt model, such as a Delco. Some generators are slightly different, so plan accordingly when connecting wires. NOTE: If it is impractical to use a car generator, you may set up this activity with a typical bicycle generator that operates off the wheel. You must do some experimentation to achieve similar results.
3. Take the front wheel off and run a threaded rod between the fork ends.
4. Use a piece of 3/4" plywood for the base.
5. Use single element standard automotive 12v headlights.
6. Make sure all lights are grounded to the generator.
7. It is best if you purchase an automotive ammeter and voltmeter to measure your students' results.
8. In lieu of the lights, you may just use the voltmeter to measure output. This will lessen your construction cost, but will not be as visual.

Answers to Technology Connections

1. Why did the pedaling get harder after each light was turned on?
   Answer: As more lights are turned on, the load increases on the generator due to the lights being placed in a parallel circuit. As each load (light) is turned on, the amount of current needed also increases, thus making the bicycle harder to pedal.

2. If a multi-speed bicycle is used, what is the effect of gearing on the output of the generator?
   Answer: As the student changed gears on, the multi-speed bicycle, he altered the mechanical advantage. As he moved up into higher gears, the pedal speed decreased, but it was harder to pedal. As the pedal speed increased to the point at which the student could physically handle it, the output should have increased.

3. What would be the mechanical advantage of this system?
   HINT: Measure the circumference of the bicycle wheel and generator pulley. What would happen if we were to change the circumference of either?
   Answer: Depending upon the diameter of the bicycle wheel and the diameter of the pulley on the generator, a different mechanical advantage would be obtained. For example, if the circumference of the bicycle wheel is 120 inches and the circumference of the pulley wheel is 6 inches, the mechanical advantage would be 120 to 6, or 20 to 1. If the circumference was changed on either, the mechanical advantage would also change.

4. Who produced the highest voltage and current readings, or the brightest lights? Was he or she the biggest person in class? Do you think that size makes a difference?
   Answer: Check the chart to see who produced the largest voltage and current readings, or relative brightness of the lamps. Not in all instances would the largest person have the highest readings.

TROUBLESHOOTING

Contributed by Thomas Barrowman

Suggested Activity Length – 1 to 2 weeks

Activity Objective – In this activity, students will learn how to identify two-cycle internal combustion engine problems. Fundamental operating principles will be learned.

Instructor Guidelines – Glow Plug Engine Troubleshooting: Air is assumed to be always present for combustion, as air readily enters the cylinder along with any fuel present.

You can use an ohmmeter to test a glow plug without unscrewing it from the cylinder. A zero reading is correct. The plug is bad if it has an infinity reading.

The reed or butterfly valve may have grit under it. This would not allow fuel to enter the cylinder. Observe the direction of the retainer spring before removing it so it can be replaced correctly. Clean the reed and the reed seat, and reassemble the valve.

Be sure the O ring between the gas tank and the engine seals properly. Fuel will not be picked up properly if the O ring is leaking.

Four rechargeable D cells hooked in parallel can be used for the glow plug battery. An electric starter for the glow plug engines facilitates engine starting. Use a 12-volt motorcycle battery to power the starter.

A supplier for glow plug engines is Hobby Shack, 18480 Bandilier Circle, Fountain Valley, CA 92728-8610.

2- or 4-cycle Engine Troubleshooting: The first item to check is the spark. Remember, the spark has to jump the plug gap at working cylinder pressure. A blue spark should jump at least 1/4". Follow any good repair manual for repairing and adjusting the carburetor. Caution students not to come in contact with the spark plug, as the spark voltage is quite high.

Answers to Technology Connections

1. By practicing on an engine, using the preceding suggestions, you will be able to attribute the engine problem to one of the three subsystems (fuel, compression, and ignition). These are the only systems that have to function to operate the engine for two minutes. Many other subsystems are present to maintain engine operation (for example, lubrication, cooling, transmission, electrical, charging).
   Answer: Ignition, fuel, and compression are the most common problems of internal combustion engines.

2. Name and describe three non-fossil fuel-burning power systems that could be used for transportation vehicles.
   Answer: Solar-powered electric motor – Solar energy is stored as chemical energy in a battery or used directly by the electric motor.
   Fuel cell – Hydrogen combines with oxygen to produce electricity and to operate an electric motor.
   Battery-operated motor – Electric motors with new types of batteries such as a sodium/sulphur or lithium/iron sulphide.

MY HERO!

Contributed by Alan Horowitz

Suggested Activity Length – 1 to 3 days

Activity Objective – One of the available sources of power is chemical energy. In this activity, students will assemble a chemically powered take-off of a simple Hero's Engine invented in the second century B.C. by Hero of Alexandria. The original engine used steam as the power source.

Instructor Guidelines – Plastic 35mm film containers can usually be obtained for free from your local Fotomat^(TM) or film developing store.

Instead of the small diameter plastic cocktail straws, the straw that comes with TV tuner or circuit board cleaner spray cans can be used.

Swivels are used to attach a fish hook to a fishing line. They are inexpensive and can be purchased in most sporting goods stores.

A hand drill and small twist drill should be used to make the holes in the film cartridge. A drill press is not recommended as it is hard to clamp the material being drilled. The processes of shearing, drilling, and adhesive fastening should be discussed.

When students use a hot glue gun and glue sticks, have them work on the piece of masonite or sheet metal. Students have a tendency to get the glue all over everything and will frequently waste the glue sticks and burn themselves. *Emphasize to students that the hot glue gets very HOT!*

If possible, do this activity outdoors! This will facilitate cleanup. Carry the bucket, fishtank, or washtub outside. Other students will be curious; don't be surprised if a crowd gathers.

Krazy Glue^(TM) or another type of instant glue can be used to adhere the straw to the fill cartridge. However, instant glues are much more dangerous to use with students and are <u>not</u> recommended. Baking soda and vinegar can be used in place of the baking powder, but is messier and smells more.

Students can help with this activity by bringing in supplies from home.

A rotary lawn sprinkler can be used to demonstrate Newton's third law of motion.

Answers to Technology Connections

1. The spinning cylinder you made is based on Hero's Engine, invented in the second century B.C. by Hero of Alexandria. The original engine used steam as the power source.

2. Where did your Hero's Engine get its power? What type of materials processing would this be considered?
   Answer: When water is added to baking powder, a chemical reaction releases carbon dioxide gas (CO_2) at a rapid rate. The rapid exhaust of this gas through the small diameter straws causes the cylinder to spin.

3. Why did the cylinder continue to spin even after the fuel was used up?
   Answer: Newton's first law of motion: If an object is at rest, it tends to stay at rest; if it is moving, it tends to keep on moving at the same speed and in the same direction (inertia).

4. If your engine was discharged in a weightless environment with no friction or air resistance, what would happen?
   Answer: Without any resistance, the engine would continue spinning unless acted upon by an outside force.

5. How is a jet plane similar to Hero's Engine?
   Answer: Newton's third law of motion: When one object exerts a force on a second object, the second object exerts an equal or opposite force upon the first. The cylinder and straws are one mass, and the exhaust gases are the other mass. In a jet plane, the body and engine of the plane are one mass, and the hot engine exhaust is the other mass.

6. People process materials by forming, separating, combining, and conditioning. We used the separating and combining processes to make Hero's Engine. Can you explain where?
   Answer: The cutting process was used when the straw was sheared in half and when the holes were drilled into the cylinder. The use of an adhesive to fasten the straws in place is a combining process.

7. Glues and adhesives are used in combining processes. Wha process would nails and screws be used for? Scissors? Straws?
   Answer: Nails and screws are types of mechanical fasteners and are examples of a combining process. Cutting material with scissors or saws is an example of separating processes.

REPULSION COIL

Contributed by Alan Horowitz

Suggested Activity Length – 2 to 4 weeks

Activity Objective – In this activity, students will explore the use of electricity and electrical induction as a source of power. Induction is used by many AC motors as a driving force.

Instructor Guidelines – It is suggested that you obtain samples of a number of different size transformers, coils, and induction motors to use as visual aids.

You will need a low-voltage, electrical power supply to use with the repulsion coil. A 6–12-volt variable AC power supply works fine. Make sure it is protected with a fuse or circuit breaker.

Copper enameled magnet wire works best for the repulsion coil, but almost any other insulated wire will also work. Make sure you use a soft iron rod and soft iron wire pieces for this activity. It will work better than steel. The aluminum center rings from a buffing wheel work well for a repulsion coil projectile. Pile three or four rings on top of each other and tape them together.

Instead of using legs on the repulsion coil base, the 3/8" – 16 N.C. bottom nut can be recessed to give the base a flat bottom.

When using the 3/8" – N.C. threading die on the soft iron rod, be sure to use plenty of cutting oil and go slowly. This is a very tedious and time-consuming step in the activity. Plan on using a lot of wire. The coil "eats" wire.

You may wish to have students do this activity in teams. After the repulsion devices are completed, they can be disassembled and the parts reused.

Answers to Technology Connections

1. Electronic circuits are made up of components. Each component has a specific function in the circuit. What is the function of the switch in the repulsion coil circuit? The coil?
   Answer: The function of the switch is to turn the repulsion coil on or off. The function of the coil is to concentrate an electromagnetic field in the soft iron rod.

2. Why is it important to use low voltage (6–12 volts AC) in this circuit? What will happen if too much voltage is applied?
   Answer: The materials used for this activity are designed to handle low voltage. If too much voltage is applied, the coil can overheat and burn out.

3. Why can this device be used on alternating current (AC) only?
   Answer: A sustained induced voltage can only be made with AC because it is necessary to have the rapid expansion and collapse of an electromagnetic field.

4. What is induction?
   Answer: When electricity flows through a wire, a magnetic field builds up around the wire. A coil of wire concentrates the magnetic field. A metallic object brought near the coil may have an electric current induced into it even if there is no direct physical contact.

AIR FLIGHT

Contributed by Margaret Rutherford

Suggested Activity Length – 5 to 10 days

Activity Objective – Students will explore the possible alternatives to the space shuttle as a means of travel in space.

Instructor Guidelines
1. Put up posters, pictures, current events, and literature on walls, doors, and bulletin boards concerning various aircraft such as airplanes, aerospace planes, and the Space Shuttle.
2. Collect magazines, literature, and books on various types of aircraft. Copy sets of material where permissible.

3. Order films or videos.
4. Collect styrofoam trays. Set up lab with materials and equipment.
5. Prepare overhead transparencies.

SUGGESTED PROCEDURES FOR PRESENTATION:
1. Review the objectives. Give a brief overview of the activity.
2. Using the Concepts and Information and films or videos to present the material to be covered.
3. Discuss safety of all tools and equipment to be used. Go over classroom procedures in use of materials and clean up.
4. Demonstrate the procedure to be followed in sketching, final drawing, pattern making, and construction of the aircraft.
5. Issue supplies and assist students as needed.
6. When all students have completed this activity, have various contests for best design, highest flight, longest flight, and most unusual type of flight.
7. Suspend from the ceiling or display the aircraft.

Answers to Technology Connections

1. What are vehicles that travel above the land and water surfaces of the earth called?
   Answer: Airplanes

2. What is a vehicle with rocket boosters capable of traveling above the earth's atmosphere carrying large payloads called?
   Answer: Space Shuttle

3. Describe what the projected national aerospace plane will be like.
   Answer: The space shuttle presently carries heavy payloads into space for commercial, military, and private industrial applications. When the national aerospace plane is completed, it will provide a less expensive and more maneuverable type of transportation to and from a distance 300 miles above the earth. Together these will further explore and develop the new frontier, space.

4. Explain the differences among the following aircraft: airplane, space shuttle, national aerospace plane.
   Answer: A. Airplanes – fly above the land and water, but these are limited to height of about 100,000 feet above the earth.
   B. Space Shuttles – powered by rocket boosters and can carry heavy loads into orbit up to 300 miles above the earth and its atmosphere.
   C. National aerospace planes – not powered by heavy rocket boosters, but could fly 300 miles above the earth's atmosphere at speeds in excess of 17,000 mph.

Resources
Suggested Software: (Consult software catalogs for sources)
Paper Airplane
Flight Simulator
Aviation and Our Environment (FAA Software)
Principles of Flight (FAA Software)

NASA Resources:
EP-192 Social Sciences and Space Exploration
NASA Spinoffs 1985, NASA Spinoffs 1986
The Sky as Your Classroom
Of Wings and Things NASA/AESP 1/87
EP-171 Limitless Horizons Careers in Aerospace
NASA Film/Video Catalog

For publications and addresses write to:
Superintendent of Documents
U.S. Government Printing Office
Washington, DC 20402

Other Resources: Free or inexpensive literature, films, or videos may be ordered by asking for listings or catalogs from the following:

Guide to FAA Publications FAA-APA-PG-6
U.S. Department of Transportation
M-442-32
Washington, DC 20591

Free FAA Aviation Education Materials 1984
Superintendent of Documents
Retail Distribution Division
Consigned Branch
8610 Cheery Lane
Laurel, MD 20707

YOU'RE ON THE RIGHT TRACK

Contributed by Alan Horowitz

Suggested Activity Length – 4 to 6 weeks

Activity Objective – In this problem-solving activity, students will design and assemble a monorail-type vehicle in order to more fully understand and appreciate the engineering and technical problems faced in the installation and construction of a full-scale monorail transportation system.

Instructor Guidelines – This is a rather difficult activity for students who are unfamiliar with many of the basic tools, materials, and machinery needed for monorail construction. You may only wish to use this activity with more advanced students.

It is recommended that the teacher construct a sample monorail car before the students begin their cars.

The 9-volt DC electric motor (Radio Shack No. 273-229) is recommended for this activity. It comes with a miniature pulley already attached. A 9-volt power supply can be used in place of the battery. Hang a wire from the ceiling of the room directly above the monorail track. Solder a 9-volt battery snap to one end of this wire and connect the other end to the 9-volt DC power source. The monorail cars can be easily connected and disconnected with the battery snaps. Watch the polarity!

The track assembly can be made into a "U" shape by bending the 1/8" x 1" band iron with a Hossfeld bender or a vise and monkey wrench. It is not easy to get a smooth bend in a flat rail track but it can be done. Use a large sheet of paper to make a full-size layout of the desired radius. Gradually bend the band iron to fit the drawing. Don't make too sharp a radius in the track or the monorail cars with a long wheel base will get stuck on the curve.

Make sure you have plenty of 10–24 NC and 1/4 x 20 NC taps available. Students have a tendency to break them very easily. Use plenty of cutting oil (or STP) during the treading process.

You will need plenty of 3" or 3-1/2" long, 1/4–20 round head machine screws with washers, nuts, and wing nuts. Also, a piece of 1/4" inside-diameter tubing (plastic or metal) slid over these long machine screws makes wonderful drive wheels. Stock up on this size tubing.

Sandpaper can be cut into 1" strips and adhered to the track with rubber cement. This will increase the traction of the monorail cars. Trace the curved track sections directly onto the sandpaper and cut out with tin snips. Avoid getting any lubricant on the monorail track. Stock up on small size rubberbands. They make great Vee belts connecting the electric motor to the drive wheel.

Answers to Technology Connections

1. Existing technological systems act together to produce new, more powerful technologies. What technologies did you use to make your monorail car?
   Answer: Sheet metal production, electricity (motors, batteries, etc.), fasteners, lubricants, fabrication machines and tools, etc.

2. High-speed monorail systems could replace the fairly slow rail system used today. What undesirable outcomes could result?

    Answer: More serious accidents at high speeds, higher maintenance costs, elimination of some traditional jobs, the destruction of homes in the path of the monorail, etc.

3. Magnetic levitation (maglev) can be used to lift a train off its track. Why does this allow the train to go faster?

    Answer: By cutting or eliminating track/wheel friction, a maglev train needs much less power to move the vehicle at high speeds.

4. The transportation systems of the future will make more and more use of computers for design and control. These systems will be much faster, safer, and more efficient than the present ones. Superconductors will mean even greater advances.

SCRAMBLER

Contributed by Thomas Barrowman

Suggested Activity Length – 3 to 4 weeks

Activity Objective – Following this activity, the student will understand some of the underlying principles that are involved in transportation, energy conversion, and systems control. Human management of the system and teamwork are integral parts of the activity.

Instructor Guidelines – Following are two ideas you could use for the steering system. The second idea could be used for the front and rear axles.

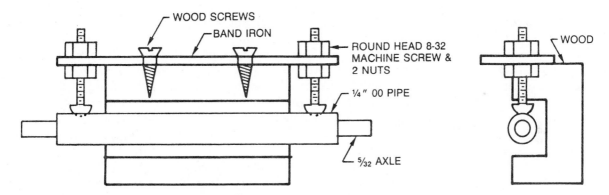

Exact front steering adjustment system

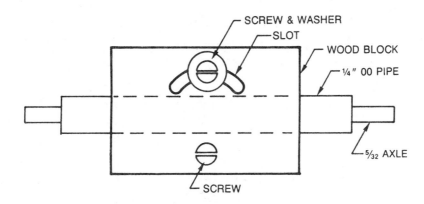

Easy front steering

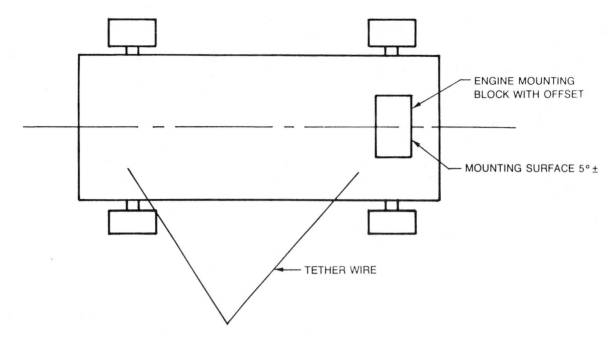

ENGINE MOUNTING
BLOCK WITH OFFSET

MOUNTING SURFACE 5°±

TETHER WIRE

Top view of vehicle

The axles can be threaded with an 8–32 die. It is important to double nut one side of each axle, as the turning wheels on one side of the car will unscrew the nuts.

The engine should be mounted on the rear of the car with the propeller close to the rear axle. The propeller will then have less movement when rough terrain is encountered. By offsetting the engine 5 degrees, the tracking performance will be improved. For better tracking, the front of the vehicle should tip slightly downward when you hold the vehicle on a string attached to the tether line.

Mount the engine on fuelproof-coated plywood or face and edge grain softwood. (Most finishes will be dissolved by the fuel. A fuel-proof finish will not dissolve.) Do not use end grain wood for mounting the engine as the screw will pull out.

Centrifugal force will force the fuel to the outside of the fuel tank. If the pickup tube is not on the correct side of the fuel tank, the engine will not receive fuel when the car is traveling in a circle. Check the instruction manual for proper placement and adjustment of the pickup tube.

The center for the tether system should weigh at least 18 pounds. Mount a 1/4" machine bolt with a washer used for holding the tether in the center of the weight. Braze an eye on the washer.

This will be used for fastening the tether. An 85-pound test fishing swivel should be used on the other end of the tether. A 12-foot, 150-pound test nylon tether line will give you a distance that will work with most vehicles. The longer the tether line, the harder it will be for the vehicle to track the circle. Keep the tether line as low as possible because a higher control line will lift the inside of the car.

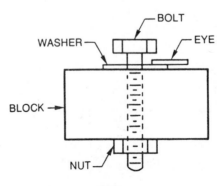

BOLT

WASHER

EYE

BLOCK

NUT

Answers to Review Questions

1. Use the following formula to calculate how fast your vehicle traveled.
   Answer: Circumference of a circle = 2πr (r = radius of a circle; π = 3.14)

$$\text{Velocity} = \frac{\text{Distance traveled in feet}}{\text{Time (in seconds)}}$$

   Example: 12 ft. radius, 2 seconds per lap

   c = 2πr
   c = 6.28 x 12 ft.
   c = 75.36 ft.

$$\text{Velocity} = \frac{75.36 \text{ ft.}}{2 \text{ sec.}} = 37.68 \text{ ft./sec.}$$

$$\text{MPH} = \frac{\text{Ft. per second x 60 x 60}}{5280}$$

$$\text{MPH} = \frac{37.68 \text{ x } 60 \text{ x } 60}{5280} \qquad \text{MPH} = 25.69$$

2. How do local speedways compensate for centrifugal force that is developed on 1/4-mile and 1/2-mile oval tracks?
   Answer: The track is banked on turns. Many cars race with the inside wheels smaller than the outside wheels.

3. Change the weight of your vehicle and time it again. How much does weight affect speed?
   Answer: The speed will increase with less weight.

4. Test the speed of the vehicle with the body removed. Does the speed change? (Remember, you are also reducing the weight of the vehicle when you remove the body.)
   Answer: Drag on such a small vehicle will not affect the performance as much as weight. In determining drag, the velocity of a vehicle is squared. Increasing speed above 40 MPH will begin to show the effects of drag.

5. Tape a white piece of paper to the wall. Start the glow plug engine and let it run with the exhaust aimed at the paper. What is the result? Why does this happen?
   Answer: The paper will become wet with unburned fuel. A two-cycle engine uses lubricating additives in the fuel. The engine uses unburned fuel to lubricate the piston. The crankshaft and bearings are lubricated with the incoming fuel and air mix.

DUST OR BUST

Contributed by Alan Horowitz

CAUTION: As we are dealing with a high-voltage static electricity charge, the chances of getting a shock with this activity are high. Normally this is not dangerous, but individuals with health problems should not attempt this activity.

Suggested Activity Length – 4 to 7 days

Activity Objective – Throughout history, people have used technology in ways that could harm the environment. In this activity, students will construct an electrostatic precipitator that filters out most particulate matter (tiny particles) from the air by using static electricity. Very large electrostatic precipitators are used by a number of industries to cut down on the pollution being emitted into the atmosphere.

Instructor Guidelines – Check with your school's science department to locate the clear plexiglass tubing and large one-hole rubber stoppers.

Also, check with your science department to locate a Van de Graff generator. Order some extra rubber bands for the generator, as they become brittle with age and break easily. A piece of brazing rod will work well in place of the 1/8" brass rod.

A continuous supply of chalk dust or smoke pumped through the electrostatic precipitator makes this demonstration more effective. A vacuum pump can be used to suck the smoke/dust into the plastic tube in place of the squeeze bulb or plastic bottle.

Brass and/or plastic lamp fittings are useful for connecting the pump and the precipitator together. A large turkey baster can also be used to inject the smoke into the precipitator. After being used a number of times, the inside of the plastic tube becomes dirty. Use a damp cloth to swab out the inside of the tube. Dry with another cloth.

Answers to Technology Connections

1. The device you have made is called an electrostatic precipitator. It filters tiny particles from the air by using static electricity.

2. What happened to the dust or smoke particles inside the tube when the brass rod was brought near the Van de Graff generator? Why?
   Answer: The smoke particles will adhere to the inside of the tube. The negatively-charged smoke particles are attracted to the positively-charged coil wrapped around the tube.

3. Would the electrostatic precipitator remove toxic (poisonous) gases from the air?
   Answer: An electrostatic precipitator removes only tiny particles that can be charged with static electricity. It will not remove toxic or any other gases. Chemical scrubbers and catalytic converters are sometimes used to remove some toxic gases.

4. Many industries send smoke and other particles into the air as a result of production. Often they can add an electrostatic precipitator to their system to help filter out these pollutants. What industries do you think would make the most use of this device?
   Answer: Power plants, chemical manufacturing, automobile industry, steel making, etc.

5. A good solution often requires a compromise. What is the trade-off (advantages and disadvantages) of adding an electrostatic precipitator to a power plant?
   Answer: Some trade-offs made when incorporating an electrostatic precipitator into a power plant or factory include the normal dangers involved when using very high-voltage electricity, and the cost of installing and maintaining the electrostatic precipitator (which is usually passed on to customers). Also, the disposal of the ash removed by the precipitator becomes a problem.

SATEL-LOON

Contributed by Thomas Barrowman

Suggested Activity Length – 2 to 3 days per group

Activity Objective – In this activity, students will gain an understanding of satellite communications. Communications, AM, FM, and CB characteristics will also be discussed.

Instructor Guidelines – The satel-loon communication activity can be done on the ground. If at all possible, do the communication activity outside and use as much distance between the units as practical.

An AM standard broadcast band transmitter can be used instead of the FM transmitter.

Tape recorders could be used to tape the outgoing signal and the signal received from the satel-loon. Comparisons between the quality of the two signals could be made.

Be sure to check the quality of reception and make fine tuning and volume adjustments before launching the satel-loon. The transmitter, satel-loon, and receiver should be placed at least 150 feet away from each other because distance will affect tuning and reception.

Parafoil kites can be purchased through Into the Wind, 2047 Broadway, Boulder, CO 80302.

Answers to Technology Connections

1. Who uses EPIRBs? Who else receives EPIRB's signal?
   Answer: Airplanes and ships use EPIRBs. Civil Air Patrol, police agencies, military, Coast Guard, and many countries will receive the EPIRB's signals. (Emergency Position Indicating Radio Beacons)

2. Compare the transmission of EPIRB signals via satellite with satellite communication using telephones or TVs.
   Answer: The EPIRB signal is transmitted up to satellites and then down to a ground station. The signal received at any one location is short and other tracking stations are alerted to help pinpoint the transmission. There are enough satellites orbiting the earth for continuous coverage for TV and telephone.

3. What type of programs are transmitted via satellite?
   Answer: News, cable, and commercial programs are transmitted via satellite.

4. What is videoconferencing? How are satellites used in videoconferencing?
   Answer: A group of people can hear and see each other with videoconferencing. Voice and a video image are sent to each party.

MAPPING THE FUTURE

Contributed by Kenneth Welty

Suggested Activity Length – 1 to 3 days

Activity Objective – Making futures wheels has been a popular learning activity with educators for years. It is a great activity for teaching students how to think about technology and the future. In addition, it helps students understand cause and effect relationships and discover how technology shapes the way we live and work.

Instructor Guidelines – Prior to implementing this learning activity, you will need to collect the materials listed in the learning activity. The composition board can be purchased in 4' x 8' sheets from your local lumberyard. You will need to cut each sheet in half. If the boards have printing on both sides, you will need to paint one side. When the learning activity is over, the boards can be reused in a similar manner to make things like storyboards, plant layouts, process flow charts, and time lines.

Introduce students to the emerging technologies listed on the learning activity. Allow students to propose their own futures wheel topics. Your students do not need to become experts on their topic to make a futures wheel. They do, however, need a basic understanding of the topic to get started.

Once they have the immediate consequence identified, their focus needs to shift from specific information to cause and effect relationships.

Consider taking your class to the library for one period. While in the library, your students could gather information about their futures wheel topics. Encourage students to use their textbook as a reference when making their futures wheel.

Using the solar energy example, demonstrate how to make a futures wheel. During your demonstration, explain how to record ideas on 3" x 5" cards, post ideas on a bulletin board, and connect ideas with string. Display your example so students can refer to it when they make their own futures wheel.

Divide your class into teams of three to four students. Have students role-play futurists working in think tanks. Allow each team to create a name for its think tank. Assign each team a work station equipped with the materials needed to make a futures wheel. Encourage students to use brainstorming techniques to generate ideas for their futures wheel. When students have finished their futures wheel, consider allowing each team to present its futures wheel to the rest of the class.

Consider putting the finished futures wheels on display so the rest of the school community can see the results of your students' forecasting activities.

To help students understand that they, to some degree, are futurists, discuss how people often describe what might happen in the future based on past trends. For example, if an athletic team is on a winning streak, they are often thought to be the favorite in future athletic events. Ask students to identify other examples.

One reason for studying the future is to plan for the future. Knowing how to make a futures wheel can come in handy when your students plan their futures. Explain how futures wheels can be used to study topics other than technology. Include among your examples topics like going to summer camp, buying a new bicycle, taking a summer job, or going to college.

People accept technological advancements as long as they serve people's needs. Discuss how technology can fulfill some needs while interfering with others. For example, automatic teller machines (ATMs) allow people to do their banking 24 hours a day. However, some people will not use an ATM because they feel a need to interact with people when they do their banking. Some people use computers to enhance their ability to process information while others are afraid of computer technology. Ask students to identify emerging technologies they think people will and will not accept in the future.

Answers to Technology Connections

The answers to these questions about futuring are not specific; there are no right and wrong answers. Encourage students to think, talk, and write about the impact of technology on their lives today and in the future.

HURRICANE ALLEY

Contributed by Alan Horowitz

Suggested Activity Length – 2 to 3 weeks

Activity Objective – In this problem-solving activity, students will explore the design and production of a wind-powered vehicle.

Instructor Guidelines – A large window fan can be used for propulsion instead of the exhaust of a heavy-duty shop vacuum.

You may want the vehicles designed by your students to have a steering mechanism of some sort. Other variations of this activity include: using popsicle sticks instead of the three 36" pieces of 1/4" x 1/4" square hardwood; using paper or cloth instead of mirrored mylar for the sail; and making the vehicle's wheel with a hole saw from 3/8" thick pine.

Use a very fine-toothed saw, such as a hacksaw, to cut the wood strips to length. A very lightweight vehicle may be difficult to control. Lightweight cars have a tendency to spin out.

A heavy wind-powered vehicle that catches a strong gust of exhaust from the vacuum will usually travel far due to the momentum it has picked up. (Inertia is the property of matter that resists change in motion. An object in motion tends to stay in motion unless acted upon by an outside force.)

Wheel lubrication and alignment are very important for maximum vehicle distance. Keep friction down to a minimum and make sure the wheels turn very freely. (Friction is a force opposing the motion of a body.)

Each wind-powered vehicle can be given a name. Students can then use a computer and a software program such as the "Print Shop" to make a sign containing the car's name and the builder's name and any other relevant information, i.e., the car's weight, the amount of materials used, best distanced traveled, etc.

Trophies can be awarded to winners in a variety of categories, such as most original, best constructed, best distance traveled, etc. Instead of individual students building the wind-powered vehicles, teams of two or three students can also work on this activity. The plastic wheels and mirrored mylar for the wind-powered vehicles can be obtained from:

Pitsco
P.O. Box 1328
Pittsburg, KS 66762
(800) 835-0686

Answer to Technology Connections

1. Can you draw a systems diagram that reflects the problem-solving approach used with the wind-powered vehicle? Label the input, process, output, monitor, feedback and comparison steps in the diagram.

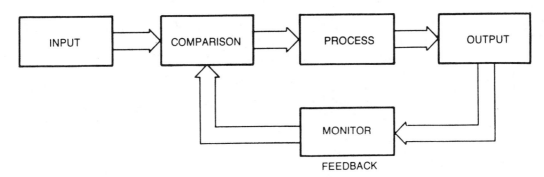

2. The design, selection, and construction of the wind-powered vehicle fall into the process step of the systems diagram. What method is used to monitor the output results?
Answer: The output results are monitored by measuring and recording the distance traveled by the vehicle. Also, a visual examination of the stability and performance of the wind-powered vehicle during testing is helpful.

3. What do we mean by feedback in the problem-solving system? How is this used to optimize the performance of the wind-powered vehicle?
Answer: Feedback is used to examine and compare the actual results (or output) of the systems model against the desired results (or input). The feedback from a failure is used to modify and optimize the vehicle's performance.

4. What is brainstorming? What is trial and error?
Answer: Brainstorming is used to generate a lot of ideas in a short period of time. A group of people trying to solve a problem together might use brainstorming. Each person in the group proposes an idea while one group member writes it down. No one in the group is allowed to criticize anyone else's idea during this process. With trial and error, many different approaches are used to solve a problem.

5. Why will the same vehicle go a greater distance if the wheels turn more freely?
Answer: Resistance to motion, caused by one surface rubbing against another, is called friction. The vehicle must overcome this friction in order to efficiently travel a greater distance. By reducing wheel friction, the travel distance will usually increase.

6. Why does the wind-powered vehicle continue to roll after the wind is removed?
Answer: The inertia of the rolling vehicle allows the vehicle to continue rolling long after the wind source is removed.

POLE POSITION

Contributed by Thomas Barrowman

Suggested Activity Length – 1 to 2 weeks

Activity Objective – In this activity, students will learn a formalized problem-solving system incorporating feedback as they modify the constructed model several times.

Instructor Guidelines – Choose the size of the wind tunnel by the size of the models you want to test. Any dimensions within reason can be used. Use any small 1/4- or 1/2-hp continuous-run air compressor, as only a small volume of air is needed. Cut a slot in the top toward the front of the viewing area and cover this with a piece of plexiglass. If wood is being used, a groove must be routed for the placement of the plexiglass. The inside surface must be perfectly smooth. Mount a fluorescent light above the opening. Weather strip a 20" x 10" plexiglass door for viewing and for placing test vehicles into the tunnel. Screw the plexiglass to a wood border so that when the door is closed the plexiglass will be even with the inside surface of the tunnel. Cut a hole in the side of the tunnel and mount the door. Drill a small hole in the side of the 2" diameter tube where the tape meets the tube. Restrict the end of a 1/4" OD soft copper pipe (1/16) and place the pipe through the hole. Bend a 3" section parallel to the sides. Secure the pipe in the center of the exit shaft with the nozzle facing the rear of the tunnel. Place a valve in the line to control the airflow. Paint the inside flat black and fill the forward end with a soda straw. You are now ready for a test.

Light a piece of punk (4th of July safety stick that is used to light fireworks) and extinguish any flame. Open the air valve, allowing a small amount of air to exit the pipe. This action forces a small amount of air through the wind tunnel. By placing the punk at different elevations you will see different airflow patterns and observe airflow characteristics eight scaled feet away from the surface. Increasing the airflow will give dramatic results. Best results are obtained with a minimum of air flowing through the tunnel.

Punk is available from the Blue Angel Company, Inc., P.O. Box 26, 129000 Columbiana-Canfield Road, Columbiana, OH 44408.

Telephone: (800) 321-9071. 100 sticks of punk is $1.09.

Answers to Technology Connections

1. Research the car companies that publish their drag form factors. Which have the best drag factors?

   Answer: The drag coefficient CD is used as a quality factor for the form of a car body. The value of CD is relatively independent of the size of the car and is between 0.30 and 0.50 on modern cars. The lower the coefficient, the less the car will disturb the air.

2. What is the purpose of each part of the wind tunnel?

   Answer: Soda straw – lines up the flow of the air transition section; helps eliminate skin drag effect and prepares for uniform entry of air into the 2" tube.

   1/4" pipe – air flowing out of the 1/4" tube causes a venturi effect and moves a larger volume of air through the wind tunnel.

3. Compare the volumes of the model cars using a water displacement method. Does the volume affect drag? How does this affect surface area?

   Answer: In most cases volume does affect drag and the drag form factor will increase with volume. Cross-sectional surface area usually increases with increased volume.

4. Why are wings used on Formula 1 cars?

   Answer: The wings are adjustable and are used to increase traction by forcing the car toward the ground.

5. Sand an airfoil shape used for airplane wings. Test this shape in the wind tunnel. Can you determine from the smoke pattern why lift is created with this shape?

   Answer: The air traveling over the upper portion has a longer path to flow to reach the back of the wing. Velocity increases, pressure above the wing decreases, and this difference in pressure produces lift.

6. What other factors influence what cars we purchase?
   Answer: Peer pressure, self image, and cost are a few.

7. Why do car companies use these same modeling techniques and similar wind tunnel tests?
   Answer: It is cost-effective to produce scaled models and test them as opposed to expensive full-scale test cars.

8. What else could you use to simulate tests and mechanical operations?
   Answer: Use a wind tunnel that has wind velocities of up to 150 mph. Use a computer for simulations.

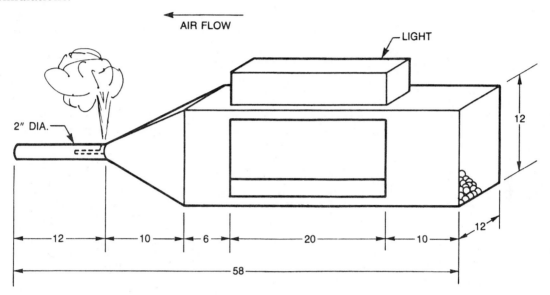

GLASSMAKING

Contributed by Thomas Barrowman

Suggested Activity Length – 2 to 3 weeks

Activity Objective – In this activity, students will learn how to process materials by formulating, mixing, and producing glass. Safety procedures and recording of formulas, times, and temperatures will be part of the activity.

Instructor Guidelines – Background information is essential for performing this activity. Obtain a copy of "Integrating Glass Technology Into Secondary School Technology Education Programs" by Michael Hacker and Paul Grey; *The Technology Teacher,* January 1985, page 9.
   Molds can be made of aluminum or steel. Press molds can be machined for a 1/2" aluminum plate. Dump molds can be formed in the top of an aluminum block.
   The annealing temperature for phosphate glass is 640 degrees Fahrenheit. Borosilicate glass is annealed at 870 degrees.
   A polariscope can be made by placing two polarizing filters, or sheets of polarizing film, at 90 degrees to each other. The glass is placed in between these two filters or sheets of film.
   The viewer inspects the glass by looking through the polarized film toward an incandescent light source. A box should be constructed to hold the light and two sheets of film or filters, with enough

space between the film to insert and test the formed glass. If bands of color or streaks of black are seen, the glass should be annealed again. An alternative batch formula follows.

Recommended Batches (Phosphate Glass)

Batch A

Formula: 20 NA_2O + .20 ZnO + .55 P_2O_5 + .05 Al_2O_3

Material	Oxide	#Moles	Formula Weight	Gram Equiv.	Weight %	# Grams in 200 gm Batch
Sodium metaphosphate ($Na_2O \bullet P_2O \bullet 4H_2O$)	Na_2O& P_2O_5	.20 .20	276	55.2	29.1	58.2
Zinc oxide (ZnO)	ZnO	.20	81.4	16.28	8.6	17.2
Ammonium phosphate ($NH_4H_2PO_4$)	P_2O_5	.35	230.1	80.54	42.5	85.0
Aluminum nitrate ($Al(NO_3)_3 \bullet 9H_2O$)	Al_2O_3	.05	750.3	37.52	19.8	39.6
				189.54	100.0%	200.0 grams

(Reprinted courtesy of *The Technology Teacher*, January 1985, Vol. 44, No. 4; published by the International Technology Education Association.)

Answers to Technology Connections

1. Bring in samples of glass. Identify the origin of the samples and what processing method was used.
   Answer: Glass is formed into the required shapes by casting, rolling, drawing, molding, blowing, or spinning. Another method is pressing, either with or without blowing.

2. Research and describe how window glass is produced. Contrast window glass with some glass produced for special purposes, such as Corning's Steuben glass.
   Answer: Ingredients for a utility glass are mixed together. Twenty to thirty percent of cullet or broken waste glass is added to this mixture. Up to 1500 tons are melted at once in tank furnaces. Special glasses for technical, scientific, or artistic uses are carefully mixed in small, carefully-controlled batches.

3. Place the operations you used in this activity in categories of forming, separating, combining, and conditioning.
   Answer: Forming – casting, pressing, spinning
               Separating – calcining, fining
               Combining – melting the mixture to form glass
               Conditioning – annealing the glass

4. Processing materials involves changing their form to make them more useful to people.

THE MEDUSA SYNDROME

Contributed by Alan Horowitz

Suggested Activity Length – 3 to 7 days

Activity Objective – In this activity, students will explore the procedures used in setting up an assembly line, engineering, and mass producing a periscope.

Instructor Guidelines – By making the periscope with two different-sized tubes, one to slide inside the other, the device can be used to look in any direction.

The mirrored plexiglass has a tendency to warp out of shape easily. Extreme care must be taken when adhering the mirrors in place, or an image distortion will result when looking through the periscope. Any diameter PVC or other rigid plastic tubing will work well as the body of the periscope.

A brace and an auger or expansive bit can be used in place of a hole saw to drill the large diameter holes in the plastic.

You may wish to discuss with students the optical principles of a periscope (reflection, refraction, magnification, etc.) and the importance of having the mirrors mounted at the correct angles. If the mirror angles are slightly off, use a belt sander to make the minor corrections. After the mirrors are glued in place, the belt sander can be used to smooth the glue joints. Be careful not to scratch or get glue on the mirrors.

The practical applications of a periscope in a submarine or tack can be discussed with students. By adding a small telescope to the eyepiece of a periscope, the image can be magnified. It is very difficult to hold both the periscope and telescope at the same time. By looking through two periscopes at the same time, it is possible to get "stereo" or three-dimensional images.

A belt sander can be used to remove the bottom of the 35mm film cartridge. The jigs and fixtures used for this activity can be simple blocks of wood clamped to the machines and used as stops. A large V block can be clamped to the drill press and used as a guide when drilling the large diameter holes in the plastic tubing.

Answers to Technology Connections

1. How does mass production and the factory system help bring prices down?
   Answer: The principles of supply and demand greatly affect prices in a free society. It is usually more efficient to mass produce items.

2. What is an assembly line? What is a conveyor belt?
   Answer: On an assembly line, each work station is assigned a specific task in the manufacture of a product. Many work stations can be combined in order to produce very complex products. A conveyor belt is one method used to transport components from one work station to another on an assembly line.

3. Why are interchangeable parts important to an assembly line?
   Answer: Interchangeable parts are important to maintain the efficiency of an assembly line. Custom parts on an assembly line will usually slow down production and increase labor costs.

4. What happens if you look through the periscope eyepiece with a small telescope?
   Answer: The image you see through the periscope will be magnified.

5. What happens if you look through two periscopes at the same time? (One for each eye, of course!) Now you know how a newt views the world.
   Answer: You will see two images. If the periscopes are perfectly aligned, it is possible to see a stereoscopic image.

SPACE COMMUNITY

Contributed by Alan Horowitz

Suggested Activity Length – 1 to 2 weeks

Activity Objective – In this activity, students will explore the engineering and construction problems involved in the design and construction of a structure for use in an alien environment.

Instructor Guidelines – Don't let students over-inflate the balloons. Over-inflated balloons will break more easily when being coated with the pariscraft. Don't use cheap balloons! Often the inexpensive variety have weak spots and don't hold up very well.

Working with pariscraft is very messy. Carefully cover your work tables with newspaper or craft paper to aid with clean-up. Students should wear lab aprons to protect their clothes from the plaster.

CAUTION: Do not let plaster or clay go down the drain of your sink! Scrape out the excess plaster from the water containers and have students wipe off their hands with a paper towel before washing them in the sink. If possible, use a sink with a plaster trap installed.

If you cut the top section off a bleach or milk bottle, it will make a good water container. Instead of using plywood for the base, masonite or very heavy cardboard can be used. Cut the pariscraft into small (4-6") pieces for complicated corners and curves. Smooth out the surface as much as possible by rubbing the softened plaster. Very little, if any, of the gauze should show if this is done properly.

Let the plaster dry thoroughly before sanding or cutting. A very fine-toothed saw can be used to cut openings in the pariscraft. A hacksaw blade cut in half and mounted in a plastic or wooden handle will make a "poor man's" keyhole saw. Files are very useful for smoothing the pariscraft's edges and surfaces.

The domed structure activity detailed in Chapter 10 can be used with this activity. The monorail system (You're on the Right Track!) activity in Chapter 15 can also be used. All sorts of models (i.e., miniature space ships, "lunar rover" vehicles, rotating radar antennas, etc.) can be constructed by students to make this an exciting activity.

Empty CO_2 cartridges can be used to simulate oxygen or fuel tanks. Empty pump-type toothpaste tubes can be used for structural members. Miniature Christmas lights can be used to add realism to the activity.

Answers to Technology Connections

1. Construction sites must be chosen to fit in with the needs of the people and the environment.

2. How does the design of your structure meet the demands of the hostile Martian environment?
   Answer: Provisions must be made for adequate food, air, and water supplies, as well as a reliable transportation system, cooling, and heating systems, electricity generation/supply, storage facilities, entertainment and recreation facilities, medical supplies and facilities, communications systems, etc.

3. What special life-support systems would probably be needed on Mars?
   Answer: Temperature, oxygen, water, food, filtration, protective clothing, and monitoring equipment (radiation, weather stations, medical, etc.) are some examples.

4. Do you think the balloon-type construction technique used in this activity could be used on a full-scale structure? Why shouldn't plaster be used for exterior walls on Earth?
   Answer: With sufficient time, resources, and the right choice of structural materials and design, it is conceivable to use balloon-type construction on a full-scale structure. Plaster is water soluble and will deteriorate quickly if used outdoors.

CYCLOID CURVE

Contributed by Thomas Barrowman

Suggested Activity Length – 2 weeks

Activity Objective – In this activity, students learn how structures are constructed by using triangular and geometric shapes for strength. Drawing and problem-solving are integral parts of the activity. Scientific principles associated with a cycloid curve are also studied.

Instructor Guidelines – A large cycloid curve ramp can be built of plywood and masonite for use with gravity vehicles. A four-foot diameter circle needs to be cut out of masonite to develop the curve. Do not try to make a scaled drawing and then expand the curve to full size. The curve will not be correct.

By saving the scrap piece of plywood that the cycloid curve is cut out of, you will have a perfectly matched divider for the ramp. This piece, cut about three inches high, can be mounted in the center of the ramp.

Hot glue can be used in place of yellow carpenter's glue for quicker assembly of some parts, such as the outer framework.

You can place another adjustable incline in front of the student's ramp to see if a marble can gain enough momentum to travel the farthest up the ramp.

Answers to Technology Connections

1. Make several different types of ramps by cutting the profile curve out of 3/4"-thick wood. Using a shaper bit, your teacher will cut a groove in the ramp for a marble. Compare the performance of these samples with the performance of the cycloid curve by rolling marbles down the ramps and trying to hit a target approximately 4 feet away. The students will usually build a ramp with a steep pitch, thinking it to be faster; but the cycloid curve will always be the fastest.

2. Could you hook a timing device to the computer and record the marble's speed or time?
   Answer: The Science Tool Kit computer program would be able to record the marble's speed and time in metric or English measurements.

3. Based on your research of structures, determine the materials that would be used to construct an amusement ride.
   Answer: Because weight is not important, steel would be the best choice. It is the easiest material to fasten permanently together through welding or riveting. Steel rusts and therefore needs continual finishing.

4. List other structures that would utilize the same construction techniques as the cycloid amusement ride.
   Answer: Ski jumps, bridges, transmission towers, and power line towers are some.

5. Sketch the shape of the foundation that would be used for a roller coaster amusement ride.
   Answer: The sketch should include large footings or driven pilings that could support the structure. Footings could also be fastened into bedrock.

HOVERCRAFT

Contributed by Thomas Barrowman

Suggested Activity Length – 2 to 3 weeks

Activity Objective – In this activity, students will understand the principles of how Hovercraft operate. Problem-solving will be used as each Hovercraft will have its own design problems.

Instructor Guidelines – Ask the students in class to bring in the directions and list of repair centers for their hair dryers. Choose a dryer that has about a 2" diameter fan in it. Contact the repair center and have the center send you one fan and motor to experiment with. Let the company know that you will be buying a few hundred of the units if they operate to your satisfaction.

Clairol, Black & Decker, Gemini, Oster, Windmere, Vidal Sassoon, Hamilton Beach, GE, Conair, Norelco, Pennys, Wards, and Sears are just a few of the companies that sell hair dryers.

If the motor will not run on 3 to 6 volts DC, it can be replaced with a DC toy motor from Radio Shack. Model Hovercraft have been constructed using everything from cooling fans for electronic equipment to 0.049 glow plug engines. By directing the air under the Hovercraft to exit at the rear of the vehicle, it is possible to use only one motor. Experiment with louvers that allow some of the air to escape and thus create forward movement.

SCAT Hovercraft, 10621 North Kendall Drive, Miami, FL 33176 (305-274-SCAT) will send you information on their Hovercraft and possibly sell you a real one for about $5,000.

Answers to Technology Connections

1. How closely does your Hovercraft resemble full-sized ones?
   Answer: Balance and airflow are very close to the real Hovercraft.

2. Hovercraft usually have one motor. How do they move ahead? How do they control direction? Can Hovercraft climb inclines?
   Answer: Some air is directed out the back of the Hovercraft. Direction is controlled by fins or louvers that direct the airflow. An incline of around 5 degrees can be climbed, depending on the size and power of the Hovercraft.

3. Air-cushion vehicles (ACVs) or Hovercraft are used to cross the English Channel. Research this area and fill in a resource chart with information relating to Hovercraft. (Ideas can be found in Chapter 15.)
   Answer: Have the students list the seven resources of technology and fill in each category with information relating to Hovercraft.

APPENDIX F GENERAL SAFETY UNIT

SAFETY RULES:

The safety rules may be copied as handouts for the students, used as posters for the lab, or made into overhead transparencies for use in the lab. Students should have a set of the safety rules for the lab and its equipment in a folder, spiral, or whatever method you use to compile a file for that student. The students should be required to have the safety rules accessible in the lab at all times.

SAFETY TEST:

Administer the safety test only after students have been properly instructed in safety. The instructor should grade the test. Depending on the standards set by your state and local board, have students retake the test if the score is below the level established. File the test as taken in the students' files or whatever system you have.

SAFETY IN TECHNOLOGY LAB ACTIVITIES

ACTIVITY OBJECTIVE

These activities are created to give students an insight to the importance of safety applications in all areas of technology, both on-the-job and off-the-job.

INSTRUCTOR GUIDELINES

Technological advancement has caused safety practices to advance to a level parallel to the complexity of materials used in production processes. Processing of new materials often generates new pollutants that are toxic and/or life threatening. Students must be made aware that wearing PPE (personal protective equipment) is as much a part of the workers' daily routine as is traditional safety rules for using hand tools and power tools. The following is a list of suggestions that would allow each student to develop appropriate safety habits, both on and off the job.

1. ORDER SAFETY CATALOGUES

   This activity will allow students to investigate many types of safety equipment and to determine the kinds of occupations that would require the use of safety equipment. In this activity, students could:

   A. Review the contents of the catalogues.

   B. List a variety of industries that would use the safety equipment.

2. DESIGN SAFETY POSTERS

   In this activity, students can use computer graphic programs such as Print Shop, Print Master, or a CAD program to create safety posters or banners.

3. HOME SAFETY AUDIT

   This activity will focus on off-the-job safety around the home and yard. Accidents off-the-job occur three to five times more often than on-the-job. Listed are some suggestions for student activities:

   A. Students take home a safety audit and check their home for safe and unsafe conditions.

B. Students read and report on safety for SAFETY EDUCATION by Wayne W. Worick, pages 91–97.

C. Students can view the video tape JOHN DEERE SAFETY PROGRAM, Order NO. DKVHU87573EN.

4. FIRE PREVENTION

Fire prevention is the only natural disaster that can be controlled and prevented if sound precautionary steps are followed. Fire can be extremely destructive and will destroy homes, families, schools, and places of employment. It is necessary for all students to have a working knowledge of fire prevention. Suggestions for lab activities include:

A. Have students identify different kinds of fire extinguishers.

B. Have students sketch a floor plan of their home and design an emergency escape route for evacuation.

C. Have students and their families go through the emergency evacuation they have planned.

5. PERSONAL PROTECTIVE EQUIPMENT (PPE)

Toxic materials can enter the body by absorption, inhalation, and ingestion. Different types of PPE have been designed for use by humans to protect against exposure to toxic materials and dangerous environmental conditions. These are suggestions for development of student awareness:

A. Briefly cover OSHA 29CFR 1900 to 1910, OSHA 29CFR 1926/1910 and OSHA 29CFR 1910 for PPE requirements.

B. Have students design a self-contained breathing apparatus to be worn in a toxic environment.

C. Have students identify and wear PPE appropriate to a technology activity.

D. Review with the students the Material Safety Data Sheets (MSDS) supplied with laboratory solvents.

6. LABORATORY SAFETY

Safety practices and habits are going to be formed by how students are allowed to act and perform their respective tasks in the technology lab. For most students, this will be their first application of safety rules and regulations. These are activities to help in the teaching of safety:

A. Have students complete a laboratory safety audit to evaluate unsafe laboratory conditions. (Use the National Standard School Shop Safety Inspection Check List published by the AVA.)

B. Give computer-generated software/safety series prior to lab activities.

C. Have students demonstrate safe operation of selected tools and equipment.

D. If possible, take the students on a field trip to an industrial site so that they can observe safe and unsafe acts and conditions.

E. Have students document accidents in the lab, if any should occur.

F. Actively involve students in safety by holding safety meetings. Have the students form a safety committee.

7. COMMUNICATIONS

To promote safety in technology, safety news should be published and displayed. Competition for accident-free man-hours among classes will serve to promote a conscious effort among students to practice safety in the technology lab. Listed here are some activities in communications which can promote lab safety:

A. Have the students write up safety news for the school newspaper.

B. Have the students generate safety reports for accident-free class periods or man-hours.

C. Give out prizes for the winners of accident-free competitions among classes.

SAFETY RESOURCES

PUBLICATIONS:

CONSTRUCTION INDUSTRY, OSHA Safety and Health Standard 29CFR 1926/1910, Revised 1987.

LABOR, OSHA Safety and Health Standards 29CFR 1900 to 1910, Revised 1987.

GENERAL INDUSTRY, OSHA Safety and Health Standards 29 CFR 1910, Revised 1983.

OSHA HANDBOOK FOR SMALL BUSINESS, OSHA 2209.

STUDENT ACCIDENT REPORTING GUIDEBOOK, National Safety Council, 425 N. Michigan Avenue, Chicago, Illinois 60611.

Worick, W. Wayne. SAFETY EDUCATION: MAN, HIS MACHINES AND HIS ENVIRONMENT. Englewood Cliffs, N.J.: Prentice-Hall Inc. 1975.

NOTE: Order OSHA publications through the OSHA area office nearest you.

CATALOGUES:

Safety Rules, Incorporated
3727 Joan Drive
Waterloo, Iowa   50702

Lab Safety Supply
P.O. Box 1368
Janesville, Wisconsin 53547-1368

The Pavlik Company
554 Green Bay Road
Kenilworth, Illinois   60043

Vallen Safety Supply Co.
10659 King William Drive
Dallas, Texas   75220

Abraxas Basic Courseware
P.O. Box 1416
Eugene, Oregon   97440

Safety Engr. & Supply Co.
P.O. Box 1626
Birmingham, Alabama 35201

FILMS AND VIDEOS:

Modern Talking Picture Service, Inc.
5000 Park Street N.
St. Petersburg, Florida
            33709-2254

Deere & Company
Distribution Serv. Center
Safety Films Department
1400 3rd Avenue
Moline, Illinois 61265-1304

GENERAL SAFETY RULES

1. Follow all safety rules at all times.
2. Never use any tools or machines unless you have passed the safety tests and have permission from the teacher.
3. Walk, never run or throw any items.
4. Report any injuries to the teacher.
5. Never cut toward your body, or point any sharp item at your body or anyone else.
6. Never play in the lab. Always concentrate on your task.
7. Never talk to anyone while you or they are operating any machine.

8. Stay out of the marked zones around machines unless you are the operator of that machine.

9. Report any damaged tools or machines to the teacher. If a machine is not operating correctly, turn it off, wait until all moving parts have stopped, then tell the teacher about the machine.

10. Use leg and arm muscles, never your back muscles, to lift heavy objects. Ask someone to help you.

11. When using sharp tools or objects, make sure your hands are not in front.

12. Check all hand tools for loose handles or any damage; report any damage to the teacher.

13. Keep your area neat. Place tools in the middle of the work area. Put scraps either in a scrap box or in the trash. Wood nails must never be put on the floor.

14. Carry only a few tools at a time.

15. Use all tools and machines in the correct way and only for the correct purposes. All safety guards must be kept in the proper position while machines are being operated.

16. Never put any objects in your mouth or ears!

17. Report any medications you are taking to the teacher.

18. Never use any tools or machines when taking medication unless authorized by the teacher.

RULES FOR SAFETY APPAREL

1. Always wear OSHA-approved safety goggles. If goggles are damaged, report the damage immediately to the teacher.

2. OSHA-approved ear protection must be worn around any machine or area where activities are occurring that are excessively loud. Check with your teacher about when and in which areas ear protection is to be worn.

3. Wear a lab apron or lab coat.

4. Sturdy, closed shoes must be worn.

5. Roll up sleeves, tuck in bows, ties, or any loose piece of clothing. Do not use tools or operate any machine when wearing any clothing that is loose.

6. Wear the proper breathing protection when working in areas where dust or fumes exist.

7. When working with materials that are sharp or can splinter, wear the proper type of gloves.

8. Remove jewelry when working in the lab.

GENERAL SAFETY TEST

TRUE/FALSE: Write the full word TRUE or FALSE in the blank to the left of the statement.

_____ 1. Sturdy, closed shoes must be worn when you are in the lab.

_____ 2. You are allowed to wear rings and bracelets that are made of gold.

_____ 3. If any of your clothing is loose, you can use tools and machines as long as you have someone watching you.

_____ 4. When lifting heavy objects, use your leg and arm muscles.

_____ 5. Walk, never run, in the lab.

FILL IN THE BLANKS: Fill in the blanks with the correct answer.

6. Where there is _____ or fumes, the proper breathing protection must be worn.

7. _____ _____ must be worn at all times in the lab.

8. _____ should be worn when moving sharp metal sheets.

9. _____ _____ must be followed at all times.

10. Always get _____ from the teacher before using any machine.

MULTIPLE CHOICE: In the blanks to the left of the questions, write the letter of the correct answer.

_____ 11. Safety glasses or goggles must be worn:
  a. while you use power tools.
  b. at all times in the lab.
  c. when there is dust in the lab.

_____ 12. You may use a power machine after you have:
  a. passed the safety test and have teacher permission.
  b. taken the safety test.
  c. used all hand tools.

_____ 13. The safety guards on the machines:
  a. must always be left on when using the machine.
  b. can be removed if the guards are in your way.
  c. must be removed before you can use the machine.

_____ 14. After you have used a machine, shut off the power and:
  a. allow the next student to use the machine.
  b. stay at machine until all moving parts have stopped.
  c. walk to the next work area.

_____ 15. You can run or play in the lab:
  a. before roll call.
  b. after the lab is cleaned.
  c. at no time.

_____ 16. When there is something wrong with a machine:
  a. turn off the power, wait for all moving parts to stop, and report the problem to the teacher.
  b. try to fix it before using it again.
  c. report this immediately, even if the machine is running.

_____ 17. If you should have any type of injury:
    a. continue working if it is minor.
    b. report it as soon as possible to the teacher.
    c. treat it yourself and go back to work.

_____ 18. You can point sharp objects at yourself or others:
    a. when you are very careful.
    b. at no time.
    c. when there seems to be no other way.

_____ 19. Safety glasses or goggles:
    a. can be any type.
    b. can be damaged when you wear them.
    c. must be OSHA-approved and in good condition.

_____ 20. Ear protection must be worn:
    a. to protect the surface of your ears.
    b. in areas where there are loud noises.
    c. only when the teacher is looking.

GENERAL SAFETY TEST ANSWERS

TRUE/FALSE:

1. TRUE
2. FALSE
3. FALSE
4. TRUE
5. TRUE

FILL IN THE BLANKS:

6. dust
7. Safety goggles
8. Gloves
9. Safety rules
10. permission

MULTIPLE CHOICE:

11. b
12. a
13. a
14. b
15. c
16. a
17. b
18. b
19. c
20. b